Oxford Textbook of Functional Anatomy

VOLUME 2

THORAX AND ABDOMEN

Oxford Textbook of Functional Anatomy

VOLUME 2

THORAX AND ABDOMEN
2ND EDITION

Pamela C. B. MacKinnon and John F. Morris

Department of Human Anatomy and Genetics, University of Oxford

With drawings by Audrey Besterman

OXFORD
UNIVERSITY PRESS

OXFORD
UNIVERSITY PRESS

Great Clarendon Street, Oxford OX2 6DP

Oxford University Press is a department of the University of Oxford.
It furthers the University's objective of excellence in research, scholarship,
and education by publishing worldwide in

Oxford New York

Auckland Cape Town Dar es Salaam Hong Kong Karachi
Kuala Lumpur Madrid Melbourne Mexico City Nairobi
New Delhi Shanghai Taipei Toronto

With offices in

Argentina Austria Brazil Chile Czech Republic France Greece
Guatemala Hungary Italy Japan Poland Portugal Singapore
South Korea Switzerland Thailand Turkey Ukraine Vietnam

Oxford is a registered trade mark of Oxford University Press
in the UK and in certain other countries

Published in the United States
by Oxford University Press Inc., New York

Library of Congress Cataloging in Publication Data
(Data available)

Typeset by Newgen Imaging Systems (P) Ltd., Chennai, India
Printed in Italy
on acid-free paper by
Grafiche Industriali

0 19 262816 X (Volume 1: Musculo-skeletal system)
0 19 262817 8 (Volume 2: Thorax and abdomen)
0 19 262818 6 (Volume 3: Head and neck)
0 19 262819 4 (Three volume set)

Preface

Medicine is changing rapidly. Medical education must therefore try to keep pace with these changes and attempt to predict the future in order to prepare doctors for their future professional lives. The trend to more scientific thinking that started in medical practice in the middle of the nineteenth century has increased, gradually gaining in momentum up to the present dramatic expansion of knowledge, both theoretical and practical, in all the sciences related to medicine. This trend, coupled with improvements in public health and nutrition, has been responsible for the marked increase in health and life expectancy which has occurred over the same period of time in many populations.

Among the many changes in medicine, few have had more impact on diagnosis than the developments in imaging the interior of the living body. This started with the discovery of X-rays in the nineteenth century and developed via the use of radiopaque substances to outline internal organs. The more recent and continuing explosion in computer technology has been harnessed to many aspects of medicine, and its application to radiological, magnetic resonance imaging (MRI), and ultrasound investigations now permits the production of sectional images of the body in any plane, which have increasingly fine spatial definition and which give ever-increasing insight into the function of the tissues being imaged. At the same time, advances in optical technology in the form of flexible fibre-optic endoscopes now permit direct visualization of the interior of hollow organs continuous with the outside world, such as the alimentary tract and airway. Similar instruments inserted through a small incision can be used to visualize the exterior of the gastrointestinal tract and of other structures in body cavities, such as the peritoneal cavity and synovial joints. Surgical procedures can also be performed by use of fine instruments passed along the endoscopes. All these new techniques have obviated the need for much of the surgery that was formerly necessary to determine a diagnosis. However, they demand, for their interpretation, a much better understanding of how the body is constructed and arranged. Sectional images cannot be interpreted unless the person trying to interpret the image has in his or her mind a really good grasp of the three-dimensional arrangement of all the tissues.

Medical training and education must, of necessity, take account of these rapid changes in medical practice.

Time in educational programmes must be found not only for new knowledge in traditional subjects, but also for the emerging disciplines such as epidemiology, medical ethics and, in particular, molecular medicine, which promises to change the face of medicine in the future. It is therefore not surprising that subjects such as topographical anatomy, which have a heavy factual content, and which at one time occupied a large part of preclinical courses, have had to be assimilated in much less time. Most programmes no longer include the time-consuming dissection of an entire cadaver, but rely instead on the use of prosected (pre-dissected) specimens, videos that can show both the normal living function and dysfunctional states, and computer-assisted learning. It is quite clear, however, that, whatever changes are made to the curricula, a sound understanding of how the body is built and functions is fundamental to all aspects of medicine, not least the interpretation of the products of the new imaging systems. Equally, it is clear that most diagnoses have to be arrived at from an examination of the intact living body and so an understanding of the functional anatomy of the living body and how it should be examined, remains a critical skill for most practitioners.

In this second edition of the *Oxford Textbook of Functional Anatomy*, we have been mindful of these continuing changes and have paid careful attention to whether or not some item of information is likely to be helpful, either in aiding the understanding of the principles of body structure or in everyday clinical practice. As a result, all guidance on dissection has been removed (good dissection guides are available and many medical schools have their own schedule); the origins and insertions of individual muscles, on which students in the past have spent a disproportionate amount of time, have also been largely removed from the text and are indicated instead in the relevant illustrations. Our emphasis on the anatomy of the *living* subject has been increased; and to underline the increasing importance of non-invasive imaging, there are more computerized tomograms and magnetic resonance images, and many of these have been merged with the text rather than forming a separate subsection. Wherever appropriate, the clinical applications of the anatomical information have been highlighted by the use of 'clinical boxes'. Questions to stimulate thought and problem-solving remain as an important part of the text, the answers are now given at the end of each chapter. The broad margin on each sheet invites the addition of personal notes which you will find invaluable in stimulating rapid recall of basic information when your memory needs refreshing in the future.

The three-volume format has been retained for ease of use, and is available as a boxed set. Volume 1 covers the musculoskeletal system; Volume 2 the thorax, abdomen and pelvis; and Volume 3 the head and neck.

Acknowledgements

It is a pleasure to record our thanks to the many people with whom we discussed various topics and who encouraged us during the preparation of the second edition of the *Oxford Textbook of Functional Anatomy*. Our special thanks are due to:

- Mrs Audrey Besterman for her anatomical drawings which are both informative and aesthetically pleasing.
- Dr Basil Shepstone and Dr Stephen Golding, who contributed the chapter on medical imaging, and who, with Dr Amit Atrey, Dr Stewart Cobb, Dr Walter Fletcher, Miss Grizelda George, Mr Michael Gillmer, Mr James Gow, Mr Lawrence Impey, Dr Derek Jewell, Dr David Lindsell, Dr Mark Monaghan, Dr Niall Moore, Mr Neil Mortensen, Dr Daniel Nolan, Dr Jeremy Price, Dr Mohammed Quttainah, Mr John Reynard, Mr William Thomson, and Dr Patrick Wheeler provided most of the clinical images for this volume.
- Glaxo Ltd for permission to use a number of the endoscopic images of the gastrointestinal tract.
- Professor Gillian Morriss-Kay who advised on the development sections.
- Our academic and technical colleagues in the Department of Human Anatomy and Genetics, University of Oxford. In particular to Professor Margaret Matthews, who provided helpful criticism of the text; Mr Roger White, the supervisor or our gross anatomy facility; Colin Beesley, for his expertise in scanning and photography; and Alicia Loreto-Gardner for cheerful support in the office.
- The many medical and surgical consultants and other colleagues in Oxford hospitals who gave up precious time to read the text and to make helpful criticisms and suggestions. In particular, we thank Mr Julian Britton, Mr Neil Mortensen, Ms Meghana Pandit, Mr John Reynard and Mr James Wilson-Macdonald, who advised on clinical aspects of various sections.
- The medical production staff of the Oxford University Press, including Miss Catherine Barnes, our patient and understanding editor, and Mr Philip Longford, who skilfully transferred the many original drawings and other illustrations into electronic format.
- All those preclinical and clinical students who have been both encouraging and thoughtfully enthusiastic, telling us candidly what they liked and disliked about various texts, including our own; in particular Deborah Home, James Gagg, and Matthew Tam, who went through the three volumes line by line.

Last, but certainly not least, we dedicate this book to all our students—past, present, and future. While searching for an appropriate form of words with which to express this, we happened on a dedicatory letter written to his students in the late eighteenth century by Sir Astley Cooper, which appears in *A Treatise on Dislocations and on Fractures of the Joints*, (4th edn), Longman, London. The idiom differs from that of the twenty-first century, but the sentiment is identical. 'This work having been composed for your use, [our] principal object will be attained if you derive advantage from it.' We hope that this will be so, not only as you study anatomy during your medical course, but throughout your subsequent careers.

Contents

Introduction to functional anatomy

Introduction to functional anatomy

Functional anatomy is the study of the structure and function of the body in its **living** state. It comprises a study of: the **skeletal system** which provides a structural framework to the body, protects vital organs and gives attachment to muscles that move us; the **muscles** and **joints** which provide for movement between our various skeletal units; the **cardiovascular system** through which oxygen and nutrients are pumped to individual cells of the body and waste materials are collected for excretion; the **lymphatic system** which is closely associated with the blood vascular system for the collection of lymph but also has a protective and immune function; the **respiratory system** through which oxygen is acquired and carbon dioxide is excreted; the **alimentary system** through which nutrients are acquired and some wastes are excreted; the **urinary system** which controls the composition of body fluids in part through the excretion of wastes; the **reproductive system** which ensures continuity of the species; the **nervous system** which receives and integrates information from both the internal and external environments and which, through controlling our speech, movements, and behaviour, enables us to respond appropriately to stimuli and to express our individual character and personality; and the **endocrine system** which, through the secretion of hormones, forms the other major control system acting in conjunction with the nervous system.

The changing form of the body and its relations to function

Always remember that each body is unique and not an assembly-line product. Each of us inherited a slightly different set of genes from our parents when we were conceived. These govern the formation of every protein from which the body is made and therefore every biochemical reaction that occurs.

The body develops in the uterus and this development usually produces a 'normal' baby. It grows further during childhood and adolescence, in part in sexually dimorphic 'growth spurts', to produce the adult form. On occasion, any part of this development may be imperfect to a greater or lesser degree.

It will be obvious to you that variation among 'normal' individuals exists and can be quite striking. It is therefore very important that you develop a concept of the **range of normality** so that you can judge what is frankly abnormal and may require attention. For this purpose, many illustrations of abnormalities are included in this book.

External differences between males and females are mostly obvious. The mature female also undergoes a monthly reproductive cycle which causes marked changes in certain internal organs. In both sexes, other more subtle changes occur throughout the day which affect the function rather than the form of the body.

Throughout life, the structure of the body responds to functional demands (for example the muscle hypertrophy that results from exercise). It also responds to abuses and injuries by repair and healing. In later adult life, ageing changes lead to senescence. Never forget that most bodies donated for examination in dissecting rooms are those of very elderly people.

The 'body' which you must consider is therefore not the static, aged form which you see on a dissecting table or as a prosection, but rather a living, dynamic organism, constantly changing and responding to the functional challenges of its environment.

Terms used in anatomical description

For ease of communication and convenience of description, the body is always considered as standing erect, facing ahead, the arms by the sides and the palms of the hands facing forward with the fingers extended (**1.1**). Place yourself in this position and note that this 'anatomical position' differs in a number of ways from your normal standing posture (see 'position of function' of upper limb, Vol. 1, p. 69).

The terms **anterior** and **posterior** refer to structures facing the front and the back of the body, respectively (**1.2**). The situation is a little different in the head and brain.

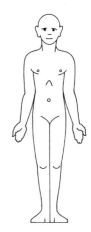

1.1 The 'anatomical position'.

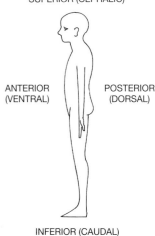

SUPERIOR (CEPHALIC)

ANTERIOR (VENTRAL)　　POSTERIOR (DORSAL)

INFERIOR (CAUDAL)

1.2 The 'anatomical position'.

Structures in the antero-posterior midline are said to be **median**; those close to the midline, **medial**; and those further away, **lateral**. Structures toward the head (above) are usually referred to as **superior** or **cephalic**, or, if they are in front and above, as **rostral**. Structures below are referred to as **inferior** or **caudal**.

Proximal means nearer to the origin of a structure; **distal** is the opposite. **Superficial** means nearer to the skin; **deep** is the opposite.

Anatomical planes (1.3)

- **Sagittal**—a vertical plane lying in the antero-posterior plane (longitudinal).
- **Coronal**—a vertical plane at right angles to the sagittal.
- **Transverse**—a horizontal plane at right angles to both coronal and sagittal; the term **axial** is used to describe this plane in CT and MRI.
- **Oblique**—any plane that is not sagittal, coronal, or transverse.

Movements (1.4)

This section concentrates on movements which occur in the trunk. For a more general list, see Vol. 1.

- **Flexion**—a forward or anterior movement of the trunk or a limb.
- **Lateral flexion**—bending of the forward facing trunk or head to either side.
- **Extension**—a backward or posterior movement.
- **Abduction**—a movement away from the midline of the body.
- **Adduction**—a movement toward the midline of the body.
- **Rotation** can occur at certain joints (e.g. in the thoracic spine).

Prone and **supine** are terms used to describe the position of the head and trunk lying, respectively, face down and face up.

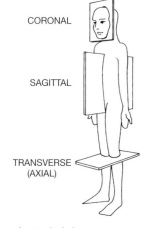

1.3 Anatomical planes.

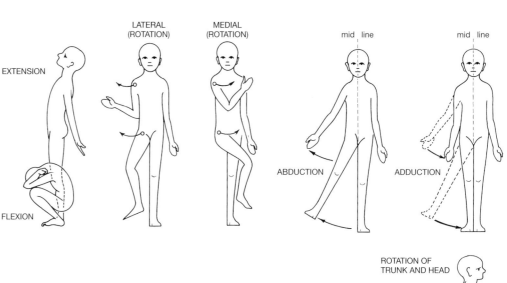

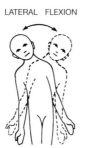

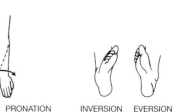

1.4 Anatomical terms used to describe movements.

Body tissues and systems

Body tissues and systems

The thorax and abdomen (torso), like any other part of the body, are made up of numerous different tissues, and these tissues are organized into various body systems. The anatomical form that these tissues take in any region is the result of their evolution to fulfil a functional role. The structure of tissues can be considered on two levels. The microscopic structure is the object of study in histology. The macroscopic (naked eye) and radiological appearance of the structure is the subject of this book.

You will find it helpful to consider, in a systematic, logical manner, general aspects of each of the tissues that make up the various systems of the body, before you proceed to the details of any part of the body,. This chapter therefore outlines some of the important topics that should be considered for each tissue; although not all will necessarily be relevant in the consideration of any single part of the body. The outline concentrates on the gross aspects of the tissues and must be supplemented by a study of their microscopical anatomy if the functioning of the tissues is to be fully understood.

Skin

The skin consists of an epithelium of keratinized stratified squamous cells (**epidermis**) on a base of connective tissue (**dermis**).
Consider:

- **Skin colour**: black, brown, or yellow skin depends primarily on the amount of melanin pigment secreted by the melanocyte cells of the epidermis. Melanin helps protect the deeper layers of the epidermis from the harmful effects of ultraviolet light.
- **Degree of keratinization**: keratinization is protective (compare the sole of the foot and the eyelid) and is increased by mechanical stress on the skin (e.g. calluses caused by heavy manual work).
- **Dermal ridges** of the hands and feet: these improve the ability of the skin to grip and assist in texture recognition (movement of ridged skin over an object produces vibrations that are sensed by cutaneous nerve endings). They are not present in truncal skin.
- **Nails** are horny plates of modified epidermis which cover the dorsum of the distal phalanges and provide a firm base for the pulp of the fingers or toes (see Vol. 1).
- **Degree of hairiness** and **type of hair**: the palms, soles, eyelids, and penis are hairless. The **density** and **coarseness** of the hair differs from region to region. The distribution of body hair is sexually dimorphic, the male pattern being dependent on circulating androgens. Abnormal hair distribution can therefore reflect endocrine imbalance.
- **Sweat glands**: There are two types:
  - **Eccrine** sweat glands (**2.1**) are present in almost the entire surface of the human body (but not the lips, eardrum, or parts of the genitalia). They are simple tubular glands with a secretory coil in the dermis and a narrow duct which spirals up through the epidermis to open on to the skin surface. In response to increases in ambient or body core temperature they secrete a colourless, watery saline. Evaporation of this secretion causes loss of heat through the skin, thus helping to regulate body temperature. Eccrine sweat glands are innervated by sympathetic (postganglionic, cholinergic) nerves.
  - **Apocrine** sweat glands are larger. In humans they are found only in the axilla, ano-genital region, and mammary areola. Their duct usually opens into a hair follicle. Their saline secretion contains organic material derived from the secretory cells and is more viscous than that from eccrine glands. When extruded on to the skin the organic material is decomposed by bacteria, generating a characteristic musky odour thought to have sex attractant (pheromone) properties. Apocrine glands develop at puberty under the influence of sex hormones and are innervated by sympathetic (postganglionic but noradrenergic) nerves.

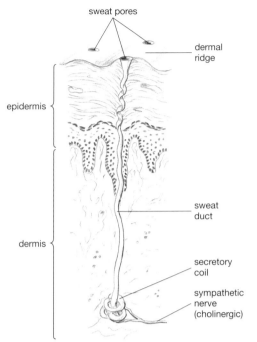

2.1 Eccrine sweat gland in the skin.

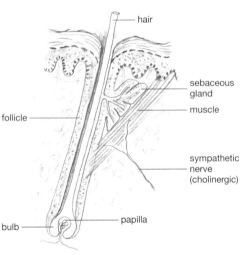

2.2 Hair follicle and sebaceous gland.

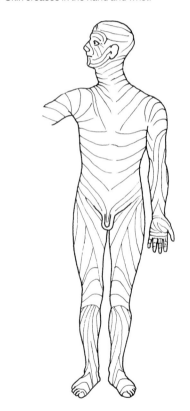

2.3 Skin creases in the hand and wrist.

2.4 Skin minimum tension lines.

• **Sebaceous glands** also open on to hair follicles (2.2) and are found over most of the body surface (not palm or sole). Their oily secretion (sebum) is derived from the disintegration of whole cells in the secretory acini (holocrine secretion). It provides a protective covering for the skin and its hairs. Sebaceous glands are not innervated, but their secretion is stimulated by androgens in both males and females, and is therefore activated at puberty. If the ducts become blocked, the secretions stagnate and can become infected, causing acne.

– The **areolar glands** of the nipple are specialized sebaceous glands.

• **Skin creases** (2.3). These are found especially around joints where the skin is firmly attached to the underlying tissues. The pronounced creases should be distinguished from the fine crease-like lines that

appear in the skin of old people and are caused by degeneration of collagen fibres and reduced attachment of the skin to underlying tissues.

• **Skin minimal tension lines** (2.4) develop, over time, as a result of skin movements caused by the contraction of underlying muscles. In general, the lines are at right-angles to the underlying muscle fibres and reflect the orientation of bundles of collagen in the dermis, running parallel to the lines. In general, the lines approximate to skin wrinkles. Skin tension lines are of considerable importance in surgery because failure to follow their orientation when making a skin incision will predispose to large, unsightly scars. Incisions made along the lines heal with fine, inconspicuous scars.

- The **blood supply of the skin** is derived from local vessels. Capillary loops in the nail bed can be seen if the skin is cleared with a drop of oil and observed with a dissecting microscope. The thermal sensitivity of skin vasculature can be demonstrated readily by immersion in hot and iced water. In cold conditions, much of the cutaneous blood supply is short-circuited from arterioles to venules by arterio-venous anastomoses.
- The **innervation of the skin**: stimuli to the skin vary in terms of their energy (mechanical or thermal), their spatial distribution, intensity, and rate of change. Many cutaneous sense organs are specialized to detect mechanical or thermal stimuli; yet others are less specific. Some respond rapidly and transiently, others in a more sustained manner.
  - Differences in packing density of the receptors produce large differences in the spatial discriminatory power of different regions. This is usually measured by the two-point discrimination threshold—the smallest distance between two simultaneously applied mechanical stimuli that can be perceived as two rather than one (see Vol. 1, Appendix).
  - For any area of skin you will need to know: the **local nerve** that supplies the area (for diagnosis of peripheral nerve injuries and for administration of local anaesthesia) and the **spinal nerve root** that supplies the area (for diagnosis of spinal cord and spinal nerve damage). The area of skin supplied by a spinal nerve is called a **dermatome**.

Fascia

Superficial fascia

Superficial fascia is the subcutaneous connective tissue which merges with the dermis of the skin. It consists of an aqueous matrix in which are various types of cell, including fat cells, fibroblasts with their bundles of collagen fibres, plasma cells, mast cells, and macrophages. These vary considerably in amount from area to area. It therefore provides a compartment which either tethers the skin to the underlying tissues or allows it to move over them; it also provides a store of energy (fat), and cells which protect against invasive organisms.

You should consider:

- The **fibrous tissue** content: this determines the attachment of the skin to deeper structures. Fibrous attachments are prominent at skin creases, and form the suspensory ligaments of the axilla, breast, and penis. In the anterior abdominal wall, a distinct fibro-elastic sheet (the membranous layer of superficial fascia) is present, especially in the lower abdomen.
- The extent of **fat** deposition and its regional variation: compare the thigh and abdomen with the eyelids, dorsum of hand, and penis. Most subcutaneous fat is 'white' adipose tissue in which adipocytes store fat in single, large droplets. Its distribution becomes sexually dimorphic at puberty; its extent is largely dependent on the balance between food intake and energy expenditure. Some 'brown' adipose tissue is found in newborn humans. Its cells contain many small fat droplets and mitochondria. It is well supplied with capillaries and sympathetic nerves and provides a rapidly available source of energy.
- The presence of **fluid-filled** fibrous sacs: these subcutaneous **bursae** allow the skin to move freely over bony prominences.
- The **superficial vessels and nerves** which pass through the fascia to reach the skin. A superficial system of veins and lymphatics runs within the superficial fascia. The nerves control the blood vessels and sweat glands, and convey information from receptors in the skin to the central nervous system.

Deep fascia

Deep fascia is a dense fibrous connective tissue that covers and ensheathes muscles and is attached to bones. In some parts the skin is strongly attached to it. It provides extra sites for muscle attachment and forms partitions (intermuscular septa) between muscle groups with different actions. In most regions it is mainly fibrous, but always contains some fat and fluid. Its thickness varies considerably. Over organs that have to expand, such as the pharynx, the deep fascia is very thin, but in areas such as the thigh it forms a non-expansile sleeve which is important in the mechanics of venous return. Many larger texts name the parts of the deep fascia according to the muscle that it covers, but only a few of these are important to remember.

The skeleton

Bone

Bone is a connective tissue that owes its great strength to a matrix containing collagen fibres on which crystals of calcium hydroxyapatite have been laid down. In most mature bone the collagen fibres are laid down by osteoblasts in osteons (Haversian systems) which consist of concentric lamellae of matrix and its cells (osteocytes), surrounding a central canal containing vessels and nerves and osteoclasts (cells which break down bone). In the earliest bone, and in bone formed immediately after fractures, the collagen fibres are randomly arranged, forming 'woven' bone.

Compact bone forms the cortex, and most of the shaft of long bones. The hollow interior of most bones is braced by **trabeculae**, bony struts arranged along lines of stress, to form trabecular or **cancellous** bone. Bone is slowly, but continuously, remodelled and can thus adapt to changing environmental stresses. Bone marrow occupies the spaces within bones. In the young, most of the marrow is haemopoietic (blood-forming) but as age progresses the marrow cavity is increasingly occupied by fat.

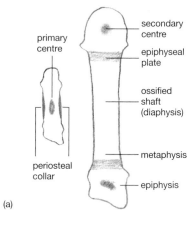

primary centre

secondary centre

epiphyseal plate

ossified shaft (diaphysis)

metaphysis

periosteal collar

epiphysis

(a)

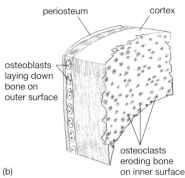

periosteum

cortex

osteoblasts laying down bone on outer surface

osteoclasts eroding bone on inner surface

(b)

2.5 The development of bones.
(a) Ossification and growth in length via epiphyseal plates; (b) appositional growth.

Development of bones (2.5)

Bones form from the mesoderm of the embryo. At first a model composed of mesenchyme (loose mesoderm) forms. In the development of many bones, this mesenchyme chondrifies to form a cartilage model of the bone, which will later ossify (**ossification in cartilage**). In the development of others, in particular the clavicle and bones of the vault of the skull (not its base), the loose mesenchyme initially condenses to form a thin, membrane-like sheet, which later ossifies (**ossification in membrane**).

The formation of bone tissue in these models starts at a **centre of ossification** (2.5a), which frequently lies centrally in the model and then spreads centrifugally. At the same time, cells of the perichondrium become osteoblasts and lay down a **periosteal collar** of bone around the developing shaft (c.f. ossification in membrane). Many bones, particularly smaller bones (such as those of the carpus, tarsus, and auditory ossicles) develop from a single **centre of ossification**. These appear over a wide period, ranging from the 6th week of intrauterine life to the 10th postnatal year, appearing later in the smaller bones.

Other bones ossify in two stages. The **primary centre of ossification** appears near the middle of the bone (in long bones, the middle of the shaft) early in development, from the 6th (clavicle)–16th week *in utero*. From this primary centre ossification proceeds towards the ends or periphery of the bone. **Secondary centres of ossification** then appear at the ends (periphery) of the bone. The first secondary centre appears just before birth (in the lower end of the femur); others continue to appear up to late teenage. The long bones of the limbs (including the metacarpals, metatarsals, and phalanges), the ribs, and the vertebrae form in this way.

While the bone develops, plates of specialized cartilage (**epiphyseal plates**) remain between the ossifying shaft (**diaphysis**) and ends of the bone (**epiphyses**). It is at these sites that linear growth of the bones continues. Linear growth is usually more marked at one end of a long bone (the '**growing end**') than at the other.

Growth gradually ceases towards the end of puberty and the epiphyseal plates become ossified so that the shaft fuses with the epiphysis. This normally occurs earlier in girls than in boys. **Fusion** of the epiphyses to the shafts is usually complete by 18–21 years.

Increase in the girth of a developing bone is the result of **appositional growth** (2.5b) in which bone is deposited by osteoblasts beneath the periosteum. At the same time, the medullary cavity is increased in size by erosion of the endosteal surface of the bone by osteoclasts. This erosion leaves behind the trabeculae that make up the cancellous bone. Whereas fusion of the epiphyses halts the growth of a bone in length, appositional growth beneath the periosteum and on the surface of trabeculae can continue throughout life, strengthening the outer surface of the bone and the trabeculae to compensate for increased mechanical stress.

During the entire growth period, the developing bone is continually remodelled to 'keep pace with' the growth. Varying tensions on bone, such as are exerted by the insertions of muscles, tendons, or ligaments, alter the contour in that area, for example the deltoid tuberosity. Abnormal tension or pressure can lead to gross deformity.

Knowledge of the times of appearance, rates of growth, and times of fusion of the secondary centres of ossification is often needed in clinical practice, for instance in assessing the skeletal age of a child in comparison with its chronological age, also in forensic medicine. These can be looked up at the appropriate time. A radiologist will often examine images of both the normal and the abnormal limb on one film in order to assess the symmetry.

Individual bones

You should be able to recognize any bone of the torso (or identify the region from which a vertebra comes) and hold it in the **position** that it occupies in the living body.

For any bone you should consider:
- The position of its **articular surfaces** and the bones with which they articulate.
- **Named parts** and **prominences**, especially those that are palpable in the living subject.
- The site of the major **muscle attachments**—these may or may not be associated roughened protrusions of the bone which give a larger surface area for attachment of the muscle.
- The site of major **ligament** and **membrane attachments**.
- The **blood supply** and position of nutrient arteries.
- The **marrow cavity** content of red or fatty marrow; its extent in children and adults.
- Any specializations of the **trabecular** pattern within the bone, or thickenings of the cortex, which reinforce particular lines of stress in the bone.
- The **ossification** of the bone; and whether ossification occurs in a **membrane** or **cartilage** model.

Cartilage

Cartilage is an avascular connective tissue which forms the articular surfaces of synovial joints, the cartilaginous models of developing bones and their epiphyseal growth plates, the pliable skeleton of the costal cartilages, nose, pinna, and larynx. Its cells (chondrocytes) lay down a resilient, very hydrated matrix which contains, in varying proportions, collagen and aggregates of glycosaminoglycans (such as hyaluronic acid) with proteoglycans. Depending on the composition of the matrix, cartilage is classified as:
- **hyaline cartilage**: this has a 'glassy'-appearing matrix and forms most articular cartilage, and the costal cartilages; it resists compression stress very well;
- **fibrocartilage**, in which the amount of collagen in the matrix (and therefore its resistance to tensile stress) is much greater;
- **elastic cartilage**, which has many elastic fibres in the matrix and is therefore very flexible (e.g. in the pinna of the ear).

Joints and their movements (2.6)

Joints are the articulations between bones. They therefore consist of different types of tissue. Their form varies widely in relation to the functional requirements of the articulation.

The degree of **mobility** varies greatly. Some joints permit virtually no movement (e.g. sacroiliac joint); at others, small gliding or angular movements occur; and at many, a large range of movement in different planes can occur.

You should consider the factors that will influence:
- the **type of movement** that can occur;
- the **range of that movement**;
- the **stability** of the joint.

In any joint, the movement that occurs will depend on two main factors: (1) the material between the two bones and the extent to which that can be deformed; and (2) the force that muscles can exert to produce that deformation.

All joints need to be relatively **stable**, but in joints that have evolved to be very mobile, stability is inevitably compromised. Therefore, in joints which have considerable mobility (e.g. the shoulder), you should consider specializations that give stability to the mobile structure. Similarly in joints (e.g. the hip) where stability is very important, you should consider specializations that optimize the mobility of the joint.

Classification of joints

Joints can be classified in a number of ways. The most common classification depends on the material that separates the bones (fibrous tissue, cartilage, or a synovial cavity). Joints may be further classified according to the type or extent of movement that can occur and whether the joint is temporary or permanent.

Fibrous joints (2.6a)
The bones are united by fibrous tissue. The extent of possible movement depends on the length of the fibrous tissue between the two bony attachments in relation to its cross-sectional area. If it is long, as in the interosseous membrane between radius and ulna or the sutures of the skull of a child at birth, then considerable movement is possible. If, however, it is short, as in the sutures of the adult skull, the peg and socket joints between the teeth and jaws, or the interosseous ligament of the inferior tibio-fibular joint, then little movement can occur.

Cartilaginous joints
In 'primary' cartilaginous joints (**2.6b**) the bones are united by hyaline cartilage. Such joints occur temporarily between the epiphyses and diaphyses of long bones and permanently in other places, such as between the first rib and the manubrium. A little flexion occurs between first rib and manubrium during respiration, but no movement should occur at epiphyseal plates, provided that the structure of the cartilage is sound (but see Vol. 1, p. 134).

In 'secondary' cartilaginous joints (2.6c) the articulating bony surfaces are covered with hyaline cartilage and united by fibrocartilage. Such joints all occur in the midline of the body: between the bodies of the pubic bones (pubic symphysis); between the bodies of the vertebrae (intervertebral discs); and between the manubrium and body of the sternum. A little movement occurs at all these joints, but endocrine-induced changes in the pubic symphysis allow greater movement during late pregnancy and parturition.

Synovial joints (2.6d)
In synovial joints the articular surfaces of the bones are covered with hyaline articular cartilage, and separated by a very thin layer of synovial fluid. These are the most common joints in the body. Many allow a considerable amount of movement between the bones, but at others virtually no movement occurs.

Mobility and stability

All joint movements can be categorized as either **sliding**, **rolling**, or **spinning** of one joint surface on the other. These fundamental motions are combined to produce the movements at a joint.

Classification of synovial joints is based on the movements that occur (determined by the shape of the articular surface):
- in **plane** joints, the articular surfaces slide over one another in a single plane;
- **hinge** and **pivot** joints have one linear axis, respectively at right angles to and in line with the main axis of the stable bone;
- a **condylar** joint is a modified hinge joint in which the curvature, and therefore the axis of the (flexion/extension) movement, varies through the movement;
- **ellipsoid** and **saddle** joints both have two axes of movement and therefore permit flexion/extension and abduction/adduction, but not true rotation;
- **ball and socket** joints can move in any axis.

  These descriptions are, of course, only approximations; all articular surfaces are ovoid to some degree. In one position the articular surfaces are most congruent (close-packed) and the joint is most stable.

For any joint you should consider:
- The **nature** and **range** of the **movements** possible. This depends on:
  - the shape of the articular surfaces;
  - the deformability of the tissues uniting the bones;
  - restrictions by ligaments;
  - restrictions by soft-tissue apposition (e.g. the arm against the trunk in adduction of the shoulder);
  - the mechanical advantage of muscles crossing the joint.

Movements at a joint can be categorized as:
Flexion–Extension
Lateral flexion
Abduction–Adduction
Medial rotation–Lateral rotation (of limbs)

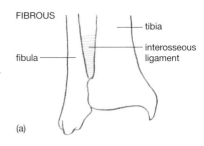

FIBROUS

(a)

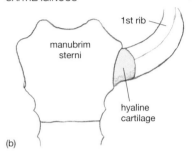

PRIMARY CARTILAGINOUS

(b)

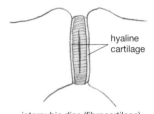

SECONDARY CARTILAGINOUS

(c) interpubic disc (fibrocartilage) showing cavity frequently present

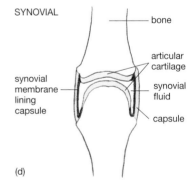

SYNOVIAL

(d)

2.6 Different types of joint.

Rotation (of head, trunk)
Pronation–Supination (of the forearm)
Inversion–Eversion (of the foot)
Protraction–Retraction (of shoulder, head, jaw)
Elevation–Depression (of shoulder, jaw)
Opposition of the thumb (and little finger)
- The **stability** of a joint. This depends on:
  - the shape of the articular surfaces;
  - the strength of the ligaments;
  - the activity of surrounding muscles

Of these, only the muscles provide **active** support. If they are paralysed, the ligaments will soon lengthen and joint deformity will ensue.

The structure of any joint reflects an evolutionary compromise between mobility and stability in relation to the function of the joint.

- The **articulating surfaces**: the bones that take part in the joint, and the shape of their articular surfaces. In some cases the area of contact in different movements is important.
- The **capsule** of the joint: its extent, attachments, strengths, and deficiencies. The capsule is fibrous and usually attached to the **margins** of the articular surfaces. Note where it deviates from the articular margins.
- The ligaments:
  - **intrinsic ligaments** are thickenings of the capsule laid down along particular lines of stress;
  - **accessory ligaments** also limit movement of the joint but are separate from the capsule.
- The **synovial cavity** and **synovial membrane**: the synovial membrane usually lines all the non-articulating surfaces within a joint. The amount of fluid in any normal joint depends on the contours of the bones within the cavity, but some larger incongruities are taken up by mobile fat pads covered with synovial membrane so that the actual volume of fluid is usually very low, and is scarcely more than a molecular layer of fluid between the articulating surfaces. The fluid has thixotropic properties (i.e. when its molecules are under pressure cross-linkages break down and the fluid becomes less viscous). On radiographs, the 'space' between the articular surfaces of the bones is occupied almost entirely by the articular cartilage, which is radiolucent.
- **Bursae**: some bursae are extensions of the synovial membrane protruding out of the joint capsule. These provide fluid-lined sacs which give friction-free movement of, for example, tendons over bones. Other bursae related to the joint are not connected with the synovial cavity.
- Intra-capsular structures:
  - **fat pads** covered with synovial membrane help to spread synovial fluid;
  - intra-articular **discs of fibrocartilage** divide the cavity of certain joints in which movements occur in two distinct axes.
- The **blood supply** to a joint: there is usually a good anastomosis of the arteries around joints which have a large range of movement, and many of the local arteries give branches to the capsule. This arrangement provides for a continuous supply distal to the joint even if the position of the joint tends to reduce the flow in some of the larger arteries.
- The **nerve supply** to a joint: the capsule of a joint has an important **sensory** nerve supply. This conveys **mechanoceptive** information to the central nervous system concerning the direction, rate, and acceleration of any movement of the joint. It also signals excessive movement via **pain** fibres.

Any nerve which supplies a muscle acting over a joint also supplies sensory fibres to that joint. More specifically, a nerve supplies that part of the capsule which is made slack by the contraction of the muscles it supplies. An intact nerve supply is thus essential to protect a joint from damage due to excessive movement.

Vasomotor fibres of the sympathetic nervous system supply arterioles of the synovial membrane.

Muscle and its actions

Muscle tissue is formed from elongated cells which contain filaments of the proteins actin and myosin, which interact to move on one another and so generate tension when intracellular calcium is increased.

Types of muscle tissue

There are three main types.
- **Skeletal** or **striated** muscle contains myofilaments arranged in parallel longitudinal bundles (myofibrils). The regularly arranged units (sarcomeres) give the muscle its microscopic striation. Skeletal muscle contracts rapidly in response to nerve stimulation; 'white' fibres contract particularly rapidly, 'red' fibres contract more slowly but are less easily fatigued and are thus more common in muscles which have a primary postural function.
- **Cardiac** muscle shares many features of striated muscle, but its cells are branched and linked both mechanically and electrically, so that the heart contracts as a co-ordinated whole.
- **Smooth** (non-striated) muscle occurs in the walls of blood vessels and of many viscera. Its single ovoid cells are very much shorter than the syncytial skeletal muscle fibres. Within the cells, the filaments of actin and myosin are not arranged in parallel sarcomeres. Rather, some of the actin is attached at various points to the plasma membrane of the cells. When intracellular calcium rises, movement of the myosin relative to the actin produces tension which can shorten the long axis of the cell. This histological difference reflects function. Smooth muscle contracts more slowly and in a more sustained manner than striated muscle.

The smooth muscle in the wall of arteries determines the resistance to flow, and in arterioles controls the blood flow into capillary beds. In the respiratory tract it regulates the diameter of the lumen and thus the resistance to air flow. Throughout the intestine it forms an inner circular and outer longitudinal muscle coat which mixes, segments, and propels the contents by peristalsis. In the urinary and gall bladder it can relax to allow urine or bile to accumulate, and contract to expel it.

Some types of smooth muscle depend less than others on their innervation; their cells form more of a functional syncytium. In the bowel wall individual muscle cells are not innervated (unitary) and have relatively few motor nerves. Much of their rhythmic contraction is myogenic and governed by pacemaker regions or by previous stretching of the muscle. Individual cells are linked by gap junctions through which a wave of electrical excitation can pass along the sheet of muscle. The innervation of such muscle enhances or depresses the rate and force of endogenous rhythmic contractions.

Circular muscle around certain parts of the alimentary and urinary tracts form **sphincters** which, controlled by autonomic nerves, can prevent the onward movement of contents. For example, the pyloric sphincter controls movement of contents from the stomach to the duodenum.

Contraction of the smooth muscle of the gall bladder and uterus is controlled much more by hormones than by nerves.

Muscles moving the skeleton

Many details of the topography and attachments (in particular the origins) of skeletal muscles are not of clinical importance, although there are obvious exceptions, such as the positions of the tendons of the wrist in trauma, and the arrangement of the muscle layers around the inguinal canal in understanding hernias. In general, it is much more important to understand the **muscle groups** that produce given movements at a joint, and their innervation.

For any muscle, you should consider:
- The **action** of a muscle in a movement. This may be classified as:
  - **agonist** or **prime mover**: this action shortens the muscle to produce the required movement;
  - **synergist**: this prevents unwanted movements which would be produced if the prime movers acted alone;
  - **antagonist**: opposes the agonists in a particular movement. During movement the antagonists are usually relaxed in proportion to the power of the prime movement;
  - **essential fixator**: clamps more proximal joints in position so that distal parts can move; for example, to produce a simple flexion of the elbow, the humerus and shoulder must be fixed.
  - **postural fixator** (e.g. of the trunk): prevents the body being toppled by movements of heavy parts which shift the centre of gravity;
  - '**paradoxical**' actions counter the force of gravity; for example, when the elbow is extended while lowering a heavy weight held in the hand, biceps generates progressively less tension, to give a controlled extension of the joint.

A particular muscle may be a prime mover in one movement, an antagonist in another, a synergist or essential fixator in others. Depending on the relationship of the attachment of a muscle to the joint over which it acts, the major component of the tension generated by the muscle may act to produce the movement, or to maintain the articular surfaces in contact while the movement occurs (2.7). Muscles with a chief role as **prime movers** therefore tend to be attached so that they have a considerable degree of mechanical advantage (e.g. biceps); muscles with a primary postural function tend to be shorter and more closely applied to a joint. The muscles of the spine illustrate this particularly well (see Vol. 1, Ch. 8).
- The **attachments** of the two ends of a skeletal muscle are to separate bones so that the muscle crosses at least one joint. Additional attachments to fibrous tissue are usually unimportant. Some muscles (especially in the face) are attached directly to the dermis of the skin.
- The **origin** of a muscle is described as its more proximal attachment, or the attachment that usually remains fixed during the prime movement produced by the muscle.
- The **insertion** is the more distal attachment and the part that usually moves.

These terms are really only used for convenience of description because muscles act differently in different movements.

An individual striated muscle fibre can contract by no more than one-third of its length. A muscle can therefore originate no closer to its insertion than the point at which the maximum movement of the joint would require around a 30% shortening of its fibres. Thus, muscles around very mobile joints must originate further away. Consider, for example, the origin of the short scapular muscles, which stabilize the very mobile shoulder joint. Their fibres originate only from the medial two-thirds of the scapular fossae and are separated from the lateral third by bursae. Fibres originating closer to the joint could not shorten sufficiently to allow the full range of medial and lateral rotation without buckling.

Some muscles appear to be attached directly to bone, but a small amount of fibrous tissue (microtendons) always intervenes. Others are attached via **tendons** (rounded bundles of fibrous tissue) or **aponeuroses** (flattened sheets of fibrous tissue). These allow: the bulk of a muscle to be separated from its point of action; the pull of a muscle to be concentrated into a small area; and the line of pull of a muscle to be altered.
- The **shape** of a muscle and the **arrangement of its fibres**. Two basic principles govern the form of a muscle. Both determine the angle at which the muscle fibres are arranged:
  - the degree of shortening possible is proportional to the length of the muscle fibres.
  - the power of a muscle is proportional to the number of muscle fibres, because each fibre can only generate a certain force.

Muscles with parallel fibres (2.8a) can shorten most but, for a given volume of muscle tissue, contain the

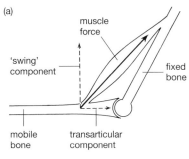

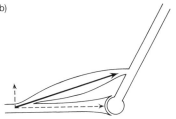

2.7 Components of muscle force at a joint in relation to their site of attachment. (a) Muscle inserted close to joint; (b) muscle inserted distant from joint.

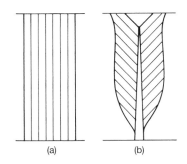

2.8 (a) Parallel and (b) pennate arrangement of muscle fasciculi.

smallest number of fibres. The number of fibres can be increased if they are oriented obliquely (pennate) to the direction of pull (**2.8b**), although this reduces the length of the fibres and thus the degree of shortening possible.

Muscles can be categorized according to their form and orientation of their fibres, but memorizing this for individual muscles is rarely necessary, although a few examples of different types and their functions should be borne in mind. For example, deltoid—a powerful abductor of the shoulder—has its main fibres arranged in oblique sets ('multipennate'), whereas psoas major—which has to accommodate the flexion and extension of the hip—has its fibres arranged in parallel.

- The **nerve supply** to muscles. You will need to know:
  - the **local nerve** supplying a muscle for the diagnosis of peripheral nerve injuries;
  - the **spinal segmental nerve** supplying the muscle for the diagnosis of spinal nerve and cord lesions.

Both of these are best learned for muscles grouped according to their main function (Volume 1: Tables 5.3.1 (p. 65), 5.4.1 (p. 73)). All muscles producing a given movement of a joint have the same segmental spinal nerve supply. For example, elbow flexors are supplied from C5 and C6 segments of the cord, and the antagonists by adjacent spinal segments (elbow extensors are supplied from C7 and C8).

Posterior primary rami of spinal nerves supply only the extensor muscles of the spine. All other muscles are supplied by their **anterior primary rami**.

Flexor and extensor muscles (the two basic embryological groups) in any limb are supplied, respectively, by **anterior** and **posterior divisions** of the anterior primary rami of spinal nerves in the plexuses supplying the limbs.

The **size of the motor units** (the number of muscle fibres supplied by one motor axon) determines the precision of action possible. For example, muscles producing finger movement have small motor units while those in gluteus maximus are very large.

- The **blood supply** to muscles. Muscles need a good blood supply, but the arteries supplying them do not need to be learned. In general, adjacent arteries supply a muscle, and a main neurovascular bundle enters the muscle at a single point. The two ends of a muscle receive supply from local vessels.

The circulatory system

The circulatory system consists of a muscular double pump—the **heart** (see Chapter 5)—linked to a system of **arteries** which branch progressively to distribute blood to the **capillary** beds in the tissues where exchange occurs. A system of **veins** returns blood from the capillaries to the heart; and a system of **lymphatic vessels** returns extracellular fluid, proteins, and particulate matter not collected by the veins to the circulation.

Two circulations are connected in series:
- a **systemic** circulation conveys oxygenated arterial blood from the left side of the heart to all the tissues of the body except the lungs, and returns the deoxygenated blood to the right side of the heart;
- a **pulmonary** circulation conveys the deoxygenated blood from the right side of the heart to the lungs, and oxygenated blood back to the left side of the heart.

The vessels in which blood and lymph circulate consist of three layers (**2.9**):
- an **endothelium** (*tunica intima*) which provides a friction-minimizing lining to the vessel and, with its basement membrane, the exchange surface of capillaries;
- a **muscle coat** (*tunica media*) which consists of elastic, fibrous, and smooth muscle tissue in varying proportions; and
- an outer coat (*tunica adventitia*) of fibrous tissue in which may run the vessels and nerves that supply the larger vessels.

Arterial supply

The **arteries** form a high-pressure distribution system. The largest arteries have a very elastic tunica media to absorb the pulsatile force of the heart contraction; smaller arteries have a more muscular coat. The **arterioles** that branch from them regulate blood flow to the tissues by contraction of the smooth muscle coat around their narrow lumen; their smooth muscle is controlled by sympathetic nerves.

For any artery, you should consider:
- Its **area of supply**.
- Its origin and parent vessel.
- Its **course**, in particular where its pulsations can be palpated, and where the vessel is exposed to injury. Arterial pulses are palpable only at certain points; usually where they cross bone. Such sites can also form useful 'pressure points' where pressure can arrest haemorrhage. The course of large vessels can

2.9 The layers of vessel walls, illustrated by an artery.

readily be mapped in the living by use of an ultrasonic Doppler-based probe. Radiologically, vessels can be demonstrated by the injection of radiopaque material into the circulation at an appropriate point (angiography).

· Its **major branches** and **mode of termination**.
· The **extent of anastomosis** with other major vessels. Some vessels, such as the central artery of the retina, are **end arteries**; some are functionally end arteries because of very limited anastomoses; yet others have plentiful anastomoses with neighbouring vessels.

Vessels of the retina can be observed directly in the fundus of the eye by use of an ophthalmoscope; capillary loops in the nail bed can be visualized if the skin is made translucent by application of oil. The capillary bed, arterio-venous anastomoses, and other small vessels of the circulatory system cannot be visualized with the naked eye.

The **degree of oxygenation** and **haemoglobin content** of the blood can best be assessed where the skin is very thin, at places such as skin creases in the palm and the conjunctiva of the eyelids.

Venous drainage

The veins form a low-pressure capacitance system. In the limbs a system of **superficial veins** is separated from the **deep veins** by the deep fascia. Like arteries, veins are lined by endothelium. In smaller veins below the heart the endothelium forms **valves** which break up the hydrostatic column of blood. The tunica media is thinner and less muscular than that of arteries because pressure in the venous system is much less.

For any vein you should consider:
· Its **origin** and the **area drained** by the vein. Remember that the origin is distal.
· Its **course**, particularly where the vein can be punctured with a needle for intravenous administration of substances or withdrawal of blood for testing; also areas which are liable to trauma.
· Any **valves** within the veins. There is a marked regional variation in the incidence of valves. They are plentiful in the limbs, particularly in the leg. Most large veins of the thorax and abdomen lack valves; negative pressure in the thorax generated by respiration sucks venous blood back to the heart, and compression by abdominal muscles also aids venous return.
· The **major tributaries** and **termination** of the vein. All veins from the limbs, head and neck, and body wall drain directly into the right atrium via **systemic** veins. Veins from the abdominal alimentary tract drain first to the liver via the hepatic **portal vein** which forms sinusoids between sheets of hepatocytes in the liver. The hepatocytes control the amount of nutrient passing from the liver via the hepatic veins to join the systemic circulation at the inferior vena cava
· The **extent of anastomosis** with other veins. In the limbs, **communicating** veins link the deep and superficial veins through the deep fascia and are important in the development of varicose veins. In the abdomen, there are small anastomoses between the portal and systemic circulations, which can become distended if pressure in the portal circulation is increased.

Lymphatic drainage

The lymphatic system is a very-low-pressure system that returns extracellular fluid, proteins, and cells to the blood system. It resembles the venous system, except that the endothelium of its blind-ending capillaries is discontinuous and therefore more permeable, and its vessels are much smaller and contain more valves.

The endothelium of lymphatic capillaries is connected by fine fibres to the surrounding connective tissue, in such a way that the presence of excess extracellular fluid pulls on the connections and opens the vessels.

The blood vascular system is protected from invasion by microorganisms via the lymphatics by the presence of **lymph nodes** along their course. These are collections of cells of the immune system, which filter the lymph and respond to foreign proteins (antigens) with an immune response. Multiple afferent lymphatic vessels enter a node, a few efferents leave via its hilum.

Lymphatic vessels cannot be detected by routine examination of a living subject unless they are inflamed or enlarged, and only the largest can readily be dissected. They can be demonstrated radiologically after injection of radiopaque dyes. Despite the difficulty in visualizing lymphatics, it is crucial that you are familiar with the lymphatic drainage of an area, since both infection and malignant tumours can spread by this route.

For any area or organ you should consider:
· The **lymph vessels** that drain the area: **superficial** lymphatics tend to run with superficial veins, **deep** lymphatics tend to run with deep vessels, particularly the arteries (the distinction between superficial and deep is the deep fascia).
· The **primary draining lymph nodes** and the extent to which these can be palpated.
· The **route** whereby lymph is returned to the bloodstream.
· The degree of **anastomosis** with other lymphatics. In general, there is very extensive anastomosis between the lymphatics serving adjacent areas, so that, when lymphatics are blocked by tumour, nodes not normally draining an area may become involved. However, in the leg, there is relatively little anastomosis between the superficial and deep lymphatics.

The nervous system

Cellular components

The nervous system acts as a unified whole, but is made up of many different specialized functional units. Each of these functional units is made up of one or more types of specialized cells. **Neurons** are the electrically excitable cells. All neurons are functionally connected with other neurons by synapses at which the connection is achieved (usually) by the release and sensing of one or more neurotransmitter chemical signals. The neurons are supported by various **glial cells**. In the central nervous sytem myelin sheaths are produced by oligodendrocytes (oligodendroglia); astrocytes (astroglia) are in intimate contact with the neurons and control their environment and influence their activity in ways that are only recently becoming clear; microglia are phagocytic, becoming particularly active after injury. In the peripheral nervous system Schwann cells fulfil all these functions.

Some neurons—**sensory neurons**—are either themselves sensitive to non-neuronal stimuli arising either outside or within the body (e.g. touch, temperature, stomach distension), or are connected to organs which transduce such stimuli (**2.10**).

Other neurons—**motor neurons**—terminate on and control the activity of non-neuronal tissue such as muscles and glands.

Any neuron that is not sensory or motor is usually referred to as an **interneuron**. Interneurons are vastly more numerous than either sensory or motor neurons. Their activity provides all the subtle processing of sensory information and memories that leads to an output of behaviour controlled by the motor neurons.

Anatomical subdivisions of the nervous system

The nervous system, which is largely bilaterally symmetrical, is divided for convenience of description into a number of different parts, each of which has its particular anatomical location and/or function.

- The **central nervous system** comprises the **brain** and **spinal cord**. It is contained within the skull and spinal canal.
- The **peripheral nervous system**, which comprises all the connections between the central nervous system and the rest of the body.

The peripheral nervous system is further divided into:

- the **somatic** nervous system (**2.10**) which (broadly) receives information from the skin and musculoskeletal system and which controls the musculoskeletal system.

- the **autonomic** nervous system (**2.11–2.13**), which controls and responds to the viscera, glands, blood vessels, and the pupil.

The peripheral nervous system is also divided into:

- the **cranial nerves**, 12 pairs of which (usually designated by Roman numerals I–XII) emerge from the brain;
- the **spinal nerves**, which emerge from the spinal cord on either side and are named according to the part of the vertebral column from which they emerge. They comprise 8 **cervical**, 12 **thoracic**, 5 **lumbar**, and 5 **sacral** nerves. Note: the first seven cervical nerves (C1–C7) emerge *above* the corresponding vertebra; C8 emerges *below* the seventh cervical vertebra, and all the remaining spinal nerves emerge *below* the corresponding vertebra.

The autonomic nervous system is further subdivided into:

- the **sympathetic** nervous system (**2.11, 2.13**), which controls many of the body's reactions to acute stress ('fight or flight');
- the **parasympathetic** nervous system (**2.12, 2.13**), which is active when body resources are being restored ('rest and recovery');
- an **enteric** nervous system of neurons intrinsic to the gastrointestinal tract.

All the nerves that make up the peripheral nervous system can be classified as either:

- **sensory** or **afferent** nerves, which convey information *toward* the central nervous system;
- **motor** or **efferent** nerves, which convey information *from* the central nervous system to the peripheral targets.

Twelve pairs of **cranial nerves** emerge from the brainstem within the cranial cavity. The first two—the olfactory (I) and optic (II) nerves—are actually extensions of the brain rather than nerves. The cranial nerves mostly supply parts of the head and neck, but the vagus nerve (X) also supplies the thoracic viscera (heart and lungs) and a large part of the gastrointestinal tract, and organs such as the liver and pancreas which develop from it.

Motor fibres in cranial nerves originate from cell bodies in the brainstem; their sensory fibres have cell bodies in ganglia associated with the nerves. Four cranial nerves [the oculomotor (III), facial (VII), glossopharyngeal (IX) and vagus (X)] also contain parasympathetic fibres which control the pupil and lens of the eye and glands in the head.

The **spinal nerves** are each formed from a **sensory** (dorsal) and a **motor** (ventral) **root**. Every spinal nerve contains both sensory and motor somatic fibres (which supply the skin and musculoskeletal system). In addition, the ventral roots of all the thoracic and the upper two lumbar nerves (i.e. T1–L2) contain sympathetic preganglionic fibres, and the last three sacral nerves (S3–5) contain parasympathetic preganglionic fibres.

All sensory (afferent) neurons have cell bodies either in the **dorsal root ganglia** of the spinal nerves or in the sensory ganglia of the various cranial nerves.

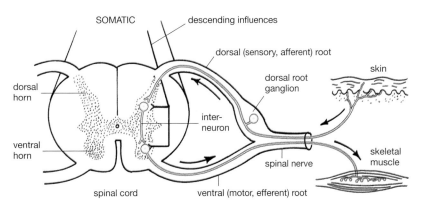

2.10 Organization of the somatic peripheral nervous system.

This is true for somatic sensory fibres from the skin and musculoskeletal system and also from the sensory fibres from the viscera, which reach the central nervous system via peripheral autonomic nerves.

Somatic **motor** (efferent) neurons originate in the ventral horn of the spinal cord and pass uninterruptedly to their target muscle fibres. Each somatic motor neuron branches to supply a number of fibres within a muscle (motor unit).

Motor (efferent) components of the autonomic nervous system always comprise a two-neuron chain linked by synapses in an **autonomic ganglion** (2.11–2.13). Thus, there are **preganglionic** sympathetic and parasympathetic neurons, which originate from cell bodies in the central nervous system, and ganglionic neurons, which give rise to **postganglionic** sympathetic and parasympathetic axons. It is the postganglionic axons which supply the target smooth muscle and secretory systems of the viscera. In the gastrointestinal tract the postganglionic fibres influence the activity of the intrinsic enteric neurons.

The motor components of the sympathetic and parasympathetic nervous system differ in the ways in which the preganglionic and postganglionic parts are arranged and, to a certain extent, in the transmitters used. In the parasympathetic system, the preganglionic fibres synapse in ganglia situated either near to (cranial ganglia) or in the walls of the viscera they supply. In the sympathetic system the synapses occur either in the **paravertebral chain of sympathetic ganglia** or in **midline ganglia** associated with the viscera and the large arteries that supply them.

Despite the restricted origin of preganglionic sympathetic neurons from T1–L2 segments of the spinal cord, the paravertebral chain of sympathetic ganglia extends from the base of the skull to the end of the sacrum, and postganglionic sympathetic fibres are distributed to all parts of the body.

Every mixed spinal nerve therefore contains nerve fibres of various types, together with their supporting **Schwann cells**, and delicate surrounding vascular connective tissue.

Each spinal nerve, on leaving the vertebral column, divides into an **anterior** and a **posterior primary ramus**. The posterior primary rami supply the extensor muscles of the spine and the skin over them; the anterior primary rami supply all the remainder, including the limbs.

For any nerve you should consider:
- The **types of fibre** that it contains. In the limbs most nerves contain sympathetic fibres and their somatic component may be motor, sensory, or mixed.
- The **origin** of the various fibres: for somatic nerves, the spinal root and peripheral nerve of origin; for autonomic nerves, the ganglion or cranial nerve of origin. Many sympathetic fibres travel as a plexus around major vessels.
- The **course** of the nerve, particularly where it is palpable or liable to trauma.
- The **major branches** of the nerve.

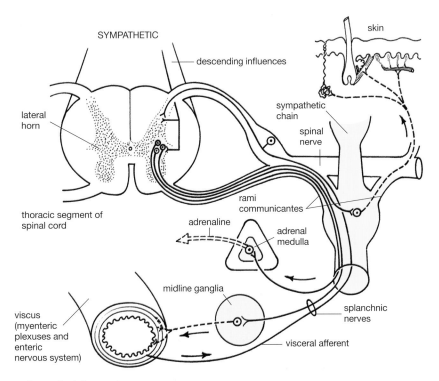

2.11 Sympathetic innervation.

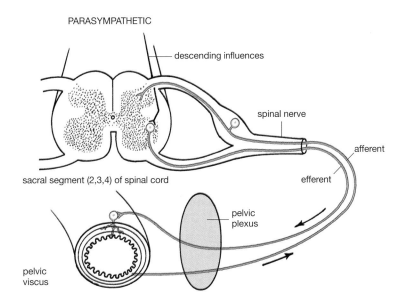

2.12 Parasympathetic innervation.

- Any **muscles** supplied by the nerve and the effect on function of damage to the nerve.
- Any **skin** supplied by the nerve and the area of anaesthesia that will be produced by damage to the nerve.
- Any **organ(s)** supplied by autonomic nerves, and their effects on the function of the organ(s). In the limbs, sympathetic fibres supply not only blood vessels, but also sweat glands and arrector pili muscles.
- The effects of **damage** to the nerve.

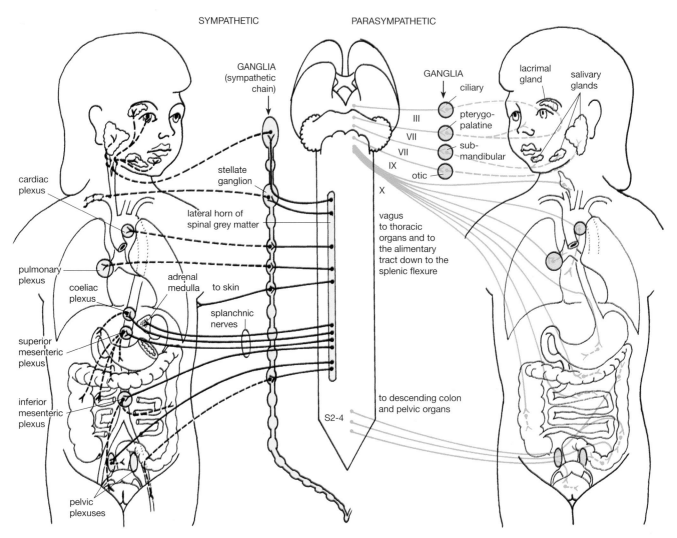

SYMPATHETIC   PARASYMPATHETIC

2.13 The organization and distribution of the sympathetic (left) and parasympathetic (right) systems.

The endocrine system (see 7.1)

In addition to control by the nervous system, all tissues receive chemical signals—hormones—produced by endocrine cells. Many of these are produced by endocrine cells grouped together to form **endocrine glands**; other endocrine cells are scattered in organs such as the gastrointestinal tract and form **diffuse endocrine systems**. Many hormones circulate widely through the body via the blood system and can affect many different tissues; others act on local cells (paracrine action). All endocrine tissues have a profuse blood supply and also receive nerves that modulate their activity.

For any endocrine tissue you should consider:
- Whether the cells are grouped as a gland or scattered.
- The **hormones** that are produced and their effects.
- The stimuli (hormonal, nervous, chemical, e.g. blood glucose) that **control** secretion of the hormones.
- The effects of too much or too little secretion

Internal organ systems

The internal organs of the body form the component parts of the various functional systems: the digestive system (gastrointestinal tract, liver, and pancreas); the urinary system (kidneys, ureters, bladder, and urethra); the reproductive system (male—testis, epididymis and vas deferens, seminal vesicles, prostate; female—ovaries, Fallopian tubes, uterus, vagina); and haemolymphoid system (spleen, lymph nodes, and vessels).

For any internal organ you should consider:
- Its principal function(s).
- The various tissue types of which it is composed.
- Its connections with other parts of the system.
- Its position in the body and principal relations.
- Its blood supply and nerve supply.
- The effects of failure of the organ's function.

Medical imaging

B. Shepstone and S. Golding

Medical imaging

B. Shepstone and S. Golding

One of the most important adjuncts to physical examination in the study of the anatomy of living subjects is medical imaging. Since the turn of the century images (radiographs) obtained with X-rays have been the standard method of visualizing many of the internal areas of the body. Images can also be obtained by use of sound waves (ultrasound imaging), or by using radiation emitted from substances which have been administered to the patient (nuclear imaging). Powerful computers are now harnessed to imaging methods to produce virtual 'slices' (often cross-sections) of the body. The first of these, computed tomography, uses conventional X-rays as the energy source. The same principles have been extended to nuclear medicine to produce emission computer-assisted tomography, and to the nuclear magnetic resonance effect to produce an image based on magnetic fields, known as magnetic resonance imaging. These techniques have proved enormously powerful in diagnosis and, in many cases, obviated the need for investigative operations. Developments in computer processing now allow the slices to be recombined to produce realistic three-dimensional images of the tissues. Ultrasound and magnetic resonance imaging (MRI) have the advantage that, unlike X-rays and other ionizing emissions, they appear not to be damaging to tissues.

Radiology

Conventional radiology

X-rays are part of the electromagnetic radiation spectrum and can be produced by bombarding a tungsten anode with electrons, using high voltages. When the electrons strike the anode, their kinetic energy is converted to heat and radiation, including X-rays. In medical radiography the tungsten anode is suspended over the patient so that the beam of X-rays passes through the body; the emerging radiation is then picked up by a detector which is usually photographic film (3.1). Since photographic film is sensitive to X-rays, the film will be exposed to a degree that depends on how much of the beam has passed through the patient and how much has been absorbed by the different tissues of the body. Air is radiolucent, bone and metal radiopaque. In the chest, good contrast is provided by the bone, soft tissues, and air-containing lungs, and a clear image of most major structures such as heart, liver, and pulmonary vessels can be obtained (3.2, H, L, and arrows, respectively). In the abdomen, however, most of the organs are of similar density with respect to X-rays and, although gas in the bowel (3.3, broad arrow) and bone can be distinguished, there is very little extra information to be obtained from soft-tissue structures, although the outlines of the kidneys are just visible (3.3, arrowheads).

Fluoroscopy ('screening')

If a fluorescent screen is substituted for photographic film, a direct image is produced. This enables movements of organs to be studied. It is customary nowadays to view the image on a television screen; an amplifier system or 'image intensifier' is usually employed which ensures a good picture with a reduction in the dose of radiation. The fluoroscopic image can then be recorded by photography or video, or by a digital computer which stores the information received (digital radiography).

Small-angle tomography

This is an old technique that has nothing to do with computed tomography except that a 'slice' is produced. It is used to overcome the superimposition effects present in the conventional radiographic image. The X-ray tube and the film move around the patient in a constant

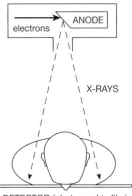

3.1 Conventional radiography. Chest radiographs are usually taken with the X-ray beam penetrating from the rear so that the heart and lungs are as close to the detector system as possible.

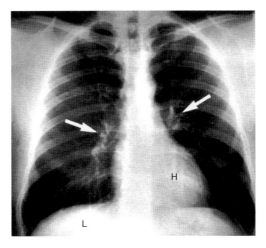

3.2 Radiograph of chest.

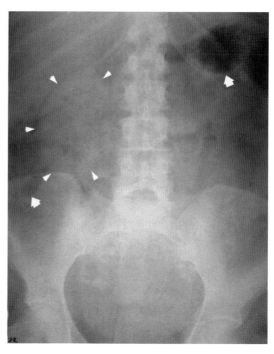

3.3 Radiograph of abdomen.

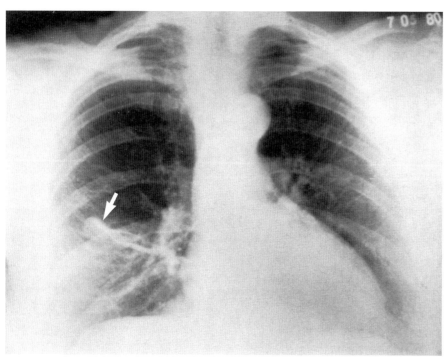

3.4 Radiograph of chest; opacity in right lung (white arrow).

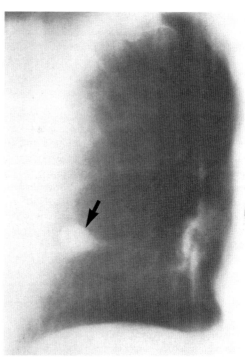

3.5 Tomogram of opacity seen in **3.4**.

relationship to the plane of interest, which therefore appears as a sharp image; the overlying and underlying areas move relative to the tube and film so that their image is blurred. The effect of tomography is therefore to produce a clear image of one plane only. It is widely used in the investigation of the chest, kidneys, and skeleton. Examine the conventional radiograph (**3.4**) which shows an opacity in the right lung (arrow). Note that in the tomogram (**3.5**) the plane of focus is such that only the nodular opacity is seen clearly.

Contrast media

It is often not possible to distinguish many organs by conventional radiography, especially in the abdomen. This problem can be overcome by the use of contrast media, which are usually either pastes of inorganic barium salts for rectal or oral ingestion, or organic substances containing iodine for intravenous administration. To demonstrate the oesophagus and stomach, a suspension of barium sulphate is given to a patient by

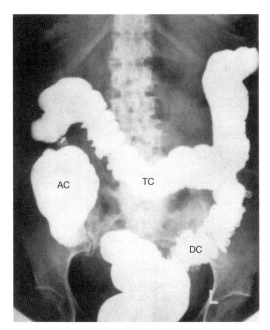

3.6 'Single contrast' radiograph of the abdomen after barium enema. The colon is filled.

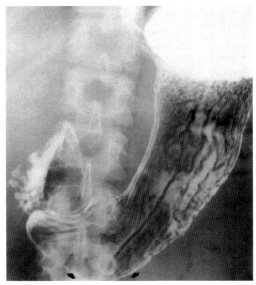

3.7 'Double contrast' radiograph of stomach. The patient has been tipped cranially causing the barium to collect in the upper part (fundus) of the stomach.

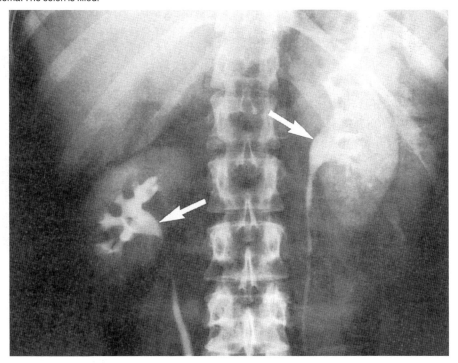

3.8 Intravenous urogram.

mouth. The colon can be similarly studied if barium is introduced through the rectum, note that in **3.6** it has filled the descending colon (DC), transverse colon (TC), and ascending colon (AC). More information is provided about the lining of the intestine if a small quantity of suspension is used and the gut is then distended with air, giving a 'double contrast' image. In **3.7** barium fills the upper part of the stomach of the recumbent patient, while the mixture of gas and barium allows detail of the lining mucosa to be seen.

Iodine contrast media have many applications; they can be injected intravenously to be excreted by the kidneys (**intravenous urogram**) (**3.8**). The contrast medium is concentrated by the renal parenchyma, making the kidney more obvious than in **3.3** and is excreted into the renal collecting system, renal pelvis, and ureter (arrows). Injections of contrast medium can also be made into a joint (**arthrogram**; see **6.3.12** of Vol. 1). In the past the method was also used to outline the bronchial tree (bronchogram) or spinal cord (myelogram) but this has now been superceded by MRI.

Contrast media are still used to study blood vessels (**angiography**). A fine tube (catheter) is inserted into an accessible vessel, for example the femoral artery,

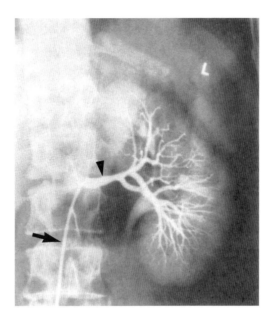

3.9 Angiogram showing the arterial tree of the left kidney.

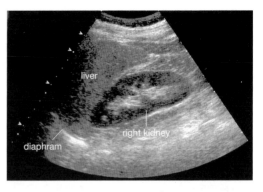

3.12 Ultrasound showing right kidney and adjacent liver.

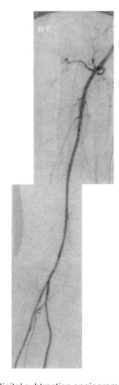

3.10 Digital subtraction angiogram of the main arteries of the arm and forearm.

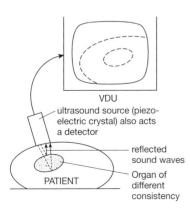

3.11 Ultrasound imaging.

and its tip manoeuvred under radiographic control into the desired vessel, where contrast media can be injected. In **3.9** the catheter (arrow) has entered the left renal artery (arrowhead) and contrast medium has outlined small vessels within the kidney. This procedure can be applied to vessels of the gut or head and neck, and even the chambers of the heart.

Digital radiography

In digital radiography each image is divided into a matrix of picture elements (pixels) and the density of each pixel is converted to a digit. The digitized image is stored on a computer and can be retrieved to a viewing screen, transmitted via telephone, and processed to improve its information content. One method of processing is **digital subtraction angiography** (**3.10**) in which images of the background are subtracted from images of the same area after injection of contrast material. It provides more information than is possible with film, and enables images of the arterial system to be produced after intravenous injection of contrast material, which was not possible with conventional techniques. It also avoids the hazards of intra-arterial injection of contrast medium, and the expense of silver-coated film.

Ultrasound imaging

Sound waves travelling in a medium are partly reflected when they hit another medium of different consistency. This produces an echo, and the time taken for the echo to reach the source of the sound indicates the distance from the reflecting surface (distance = time × velocity). Ultrasound imaging detects and analyses these echos. Ian Donald, Professor of Obstetrics in Glasgow, realized the potential of this technique and developed it to study the pregnant uterus and the fetus.

Ultrasound consists of high-frequency sound waves that cannot be detected by the human ear. When ultrasound travels in human tissue it undergoes partial reflection at tissue boundaries; thus a proportion of the sound waves return as an echo while the rest continue (**3.11**). Bone almost totally absorbs the sound and therefore no signal can be obtained from bone or the structures beyond. Neither can a signal be obtained from gas-containing viscera. These two facts are responsible for significant limitations to the use of this technique. However, it has the great advantage that there are no damaging effects at the sound energies used and it can therefore be used to monitor the developing fetus (see p. 197).

Ultrasound images are sectional. **3.12** shows the right kidney and adjacent liver and diaphragm. Vessel walls in the centre of the kidney give a bright signal; the liver tissue appears very uniform apart from at its edges. In addition to its use in obstetrics, ultrasound is used in the investigation of the heart, urogenital system, liver and biliary system, and in all joints. All current machines have a rapidly repeated scanning action, which enables movement of various organs to be studied in 'real-time'. In cardiology, for example, the movement of heart valves and chambers of the heart can be visualized (see **4.5.20**) as the heart beats.

Doppler colour-flow imaging

When a wave motion is radiated from a moving source, there is a change in frequency of the wave—the Doppler effect. This principle can be used to study moving structures, such as blood flowing through peripheral vessels. Probes emit ultrasonic waves of a given frequency. The signal received back from the body by the transducer contains the emitted frequency (due to waves reflected from stationary sources) and Doppler-shifted frequencies reflected from moving structures. The Doppler-shifted frequencies (**3.13b**) can then be filtered out, sorted for the extent of the shift, and recorded. The information can then be presented on a colour monitor with the vessels represented in different colours. An initial colour-flow image (**3.13a**) is used to locate the vessel being studied (in this case, the femoral artery) and, from the colour, to determine the direction of flow. Graphic presentation shows the speed of forward flow during systole and the small backward flow during diastole. This technique is often used to evaluate conditions in which flow is restricted (stenoses) or increased (tumour vessels).

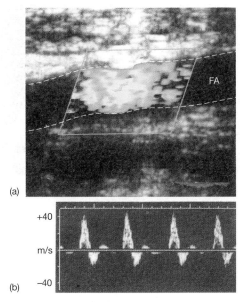

(a)

(b)

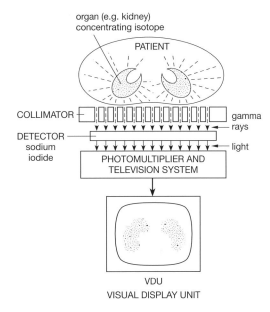

organ (e.g. kidney)
concentrating isotope

PATIENT

COLLIMATOR — gamma rays

DETECTOR
sodium
iodide — light

PHOTOMULTIPLIER AND
TELEVISION SYSTEM

VDU
VISUAL DISPLAY UNIT

3.13 (a) Doppler colour flow image of normal femoral artery (FA); (b) Doppler frequency signals.

3.14 Nuclear imaging.

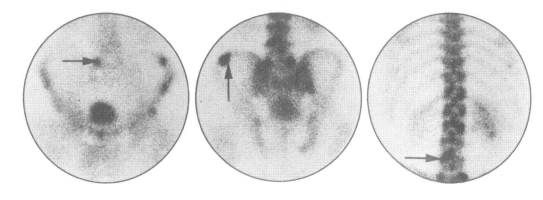

3.15 Nuclear images of pelvis and spine, showing increased uptake of isotope consistent with secondary tumour deposits (arrows).

Nuclear imaging

Nuclear medicine may be defined broadly as the use of radioactive isotopes ('radionuclides') in the diagnosis and treatment of disease. Interest to the student of anatomy arises from that branch of nuclear medicine called nuclear imaging or 'scintigraphy', whereby a 'map' of the uptake of radionuclide in a given organ system or pathological lesion can be produced by means of an instrument called a gamma camera. Analysis of such an image (which depends not only on morphology, but also on function) is then of use to the diagnostician in elucidating the nature of the patient's problem.

The basis of the gamma camera is a large, flat crystal of sodium iodide, which converts into light rays the gamma rays emitted from radionuclides which have been injected into the patient. These light rays strike a photosensitive surface and cause the emission of electrons. The latter are amplified by a photomultiplier and eventually electrical pulses are formed. These electrical pulses can be used to produce an image on a television screen or be used as input to a computer system (**3.14**).

A number of radionuclides are used in a form described as 'radiopharmaceuticals'. The principal radionuclide used is technetium-99m (which is very safe for the patient in terms of absorbed radiation dose, and has a convenient half-life of 6 hours), but others are also available, e.g. iodine-123, thallium-201, gallium-67, indium-111. These radionuclides are coupled to various compounds designed to deposit them in the organ of interest. For example, technetium-99m coupled to sulphur colloid will be phagocytosed by macrophages and so be deposited in the reticuloendothelial system. Technetium-99m phosphates will deposit in bones. In **3.15** note the focal areas of increased isotope uptake in the spine and pelvis which are due to tumour deposits from breast cancer. Iodine-123 as sodium iodide will be taken up by the thyroid. The normal uptake of such a radiopharmaceutical in an organ must be familiar to the radiologist, so that any deviation from the normal pattern is detected rapidly.

Sectional images (emission computed tomography; ECAT) can also be produced using methods very similar to those for emission computed tomography (see below). There are two varieties of this technique, known

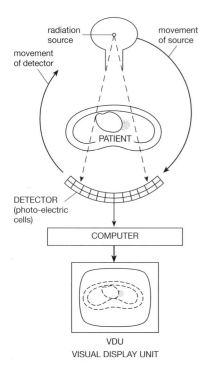

3.16 Computed tomography.

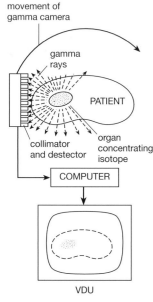

3.18 Emission computer-assisted tomography.

as single-photon emission tomography (SPET) in which solitary gamma rays (say from technetium-99m) are used, and positron emission tomography (PET) in which positrons (positive electrons) are produced (for example from a cyclotron-produced radionuclide such as fluorine-18-deoxyglucose). Both methods produce functional cross-sectional images and are widely used to investigate the brain and the heart.

Computed sectional imaging

A number of techniques can now be used to produce sectional images of the body.

It is a convention that all such transverse sectional images are presented as if the section is viewed from the feet of the patient. Thus, structures on the right of the image are on the left hand side of the patient and vice versa.

Transmission computed tomography

Computed tomography (CT) is similar to conventional radiography in that a beam of X-rays is passed through the body and is measured after it emerges. The differences between CT and conventional X-ray methods are that a very narrow beam is used and an array of highly sensitive photoelectric cells is substituted for photographic film. The beam is rotated around the patient and density measurements are made from many different angles. The data are analysed by a computer and the whole image is displayed on a TV monitor (**3.16**), dense structures such as bone conventionally being shown at the white end of the spectrum.

CT images are usually true axial sections, although coronal sections can be taken of the head by positioning the patient appropriately. However, computer programs are available whereby a sagittal or coronal image can be built up from data derived from successive axial images. CT produces extremely clear and finely detailed cross-sectional radiographs without any superimposition of surrounding structures. Thus it is possible to see organs or small differences in density caused by disease, which are almost impossible to demonstrate by conventional radiography.

Instead of producing discrete cross-sections, a spiral motion of the gantry is now used. This produces continuous cross-sections (**spiral CT**) and also operates much faster. **3.17** is an axial (transverse) CT of the abdomen in which the kidneys (K), liver (L), pancreas (P), and small bowel (B) are all clearly seen because they are outlined by body fat.

Emission computer-assisted tomography (ECAT)

This is based on the same principle as transmission computed tomography described above, except that it depends on gamma rays emitted from a radionuclide concentrated in an organ rather than on a transmitted X-ray beam (**3.18**). In **3.19**, technetium-labelled

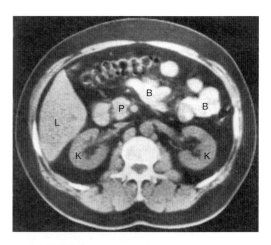

3.17 Axial CT of abdomen.

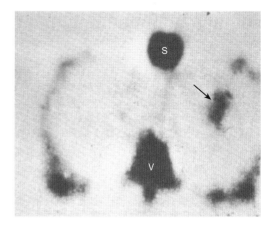

3.19 ECAT image (transverse section) of the chest, showing increased uptake of isotope in heart muscle (arrow) consistent with tissue damage (infarction).

methylene diphosphonate has been taken up in the skeleton of the chest in the same way as in the examination shown in **3.15**. However, this is a cross-sectional image, so that you see the nuclide activity in the vertebral body posteriorly (V) and the sternum anteriorly (S). The uptake within the left side of the thorax (arrow) is in an area of muscle infarction in the heart. This would not have been seen on a conventional scintigram as it would have been obscured by activity in the overlying ribs.

Magnetic resonance imaging (MRI)

MRI is a diagnostic technique based on radio signals emitted from resonating atoms within the body. The nuclear magnetic resonance (NMR) effect depends upon the fact that atoms carry a charge and can therefore be regarded as small magnets. When a large magnetic field is applied across a body there is a tendency for the nuclei to line up with this field. Nuclei spin and, rather like spinning tops, can be displaced from their axis by a field applied at an angle to the main one. If the second field is of an appropriate frequency, the atoms can be made to resonate. Since the atomic nuclei carry a charge, any movement sets up radio signals which can be detected outside the body. The interest of magnetic resonance centres on the fact that the behaviour of the resonating nuclei depends upon the atoms surrounding them, in other words their chemical environment. Thus the radio signals measured outside the patient can reflect the biochemistry of the area under examination.

The magnetic resonance scanner looks rather like a CT machine, in that the patient lies on a couch and enters a gantry which, in this case, is a very large electromagnet. Sections are taken through the area being examined and the resulting MR images are superficially similar to CT. However, CT is based on the **density of tissues** to X-rays whereas MR images reflect some aspects of the **biochemical composition of the tissues**.

The sequence of magnetic pulses can be altered (T_1- and T_2-weighting) to vary the signal from different tissues but, in both, fat (including fatty bone marrow) produces a high signal intensity (white), muscle produces an intermediate signal, and cortical bone and fibrous tissue produce low-intensity signals (black). T_2-weighting produces a high-intensity signal from fluid, which gives a low intensity signal with T_1-weighting. In 3.20 the axial section through the abdomen shows aorta (A), right and left kidneys (RK, LK), liver (L), spleen (S), stomach (St), splenic flexure of colon (C) and gall bladder (G).

Functional magnetic resonance imaging

This is a variation on the magnetic resonance principle, whereby increased brain blood flow in response to a motor or sensory stimulus can be detected and superimposed on a routine MRI of the brain. Even the desire to move a finger can now be mapped on the brain, as well as the origin of the motor action!

Three-dimensional reconstruction

The development of computer techniques and spiral CT now allows the sectional images to be combined to reconstruct the **three-dimensional shape** of the tissues being imaged and to display these on a screen. This has proved very useful, for example, in reconstructive surgery.

In **3.21** a CT data set from the chest of a patient who complained of difficulty with swallowing has been used to reconstruct the major vessels arising from the aorta and pulmonary trunk. The pulmonary trunk appears normal, but the branches of the aorta show a developmental abnormality (see p. 68). The right subclavian artery arises, not from a brachiocephalic trunk with the right common carotid artery, but as the last branch from the arch of the aorta. It then winds to the right, behind the trachea and oesophagus (which are not visualized) and creates a constriction in the oesophagus, which explains the difficulty with swallowing.

3.22 shows a reconstruction of major branches of the abdominal aorta.

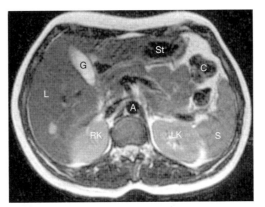

3.20 MRI of abdomen; axial (horizontal) section showing aorta (A), right and left kidneys (RK, LK), liver (L), spleen (S), stomach (St), splenic flexure of colon (C), and gall bladder (G).

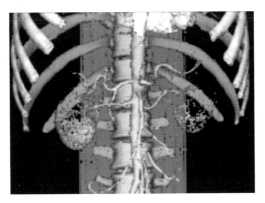

3.22 Three-dimensional reconstruction of branches of the abdominal aorta. The spine and lower ribs are shown to provide context. The coeliac, superior mesenteric, and renal arteries are shown, with parts of both kidneys.

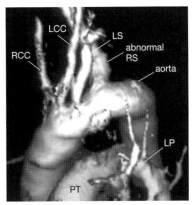

3.21 Three-dimensional reconstruction, showing the abnormal origin of the right subclavian artery (abnormal RS) as the last major branch from the arch of the aorta. It arises after the right and left common carotid arteries (RCC, LCC) and the left subclavian artery (LS). The pulmonary trunk (PT) and left pulmonary artery (LP) are visualized; the right pulmonary artery is hidden behind the ascending aorta.

Development of the body cavities

Development of the body cavities

Within the trunk, before any of the different organ systems develop, a primitive body cavity (embryonic coelom) is formed by the splitting of the intra-embryonic mesoderm. The split occurs in the midline, rostral to the prochordal plate (where the heart will start to form) and extends posteriorly into the lateral plate mesoderm on either side, forming a parietal layer which lines the ectoderm, and a splanchnic layer which covers the endoderm. The parietal layer will form the dermis, connective tissue, and serous membrane of the body wall; the splanchnic layer will form the connective tissue, muscle, and serous membranes of the alimentary, respiratory, and genitourinary tracts. Development of each individual system is presented with the anatomy of that system in the sections that follow. Development of the body cavities—the pericardial cavity around the heart, the pleural cavities around the lungs, and the peritoneal cavity in the abdomen and pelvis—are considered here, together with the development of the diaphragm, which separates the thoracic from the peritoneal cavities.

By the fourth week of fetal life the intra-embryonic coelom (or body cavity) has become a horseshoe-shaped channel. Its central curved portion, lying rostrally, will form the pericardial cavity, while the lateral extensions passing down either side of the embryo will form the pleural and peritoneal cavities. The caudal portion of each lateral extension is continuous with the extraembryonic coelom at the lateral edge of the embryonic disc (4.1).

The rapidly growing embryo folds both ventrally and transversely. Ventral folding of the head end thrusts the **pericardial cavity** ventral to the foregut, so that the cavity comes to open dorsolaterally into the two **pericardio-peritoneal canals** which, as the lung buds expand into them, will become the pleural cavities (4.3 and see 5.3.1). Transverse folding of the lateral body wall causes the right and left limbs of the intra-embryonic coelom (and also the amniotic sac) to approach each other around the ventral aspect of the embryo (4.2a). This leads to the formation of a single large **peritoneal cavity** from the intra- (and extra-) embryonic coeloms. The cavity extends from the caudal

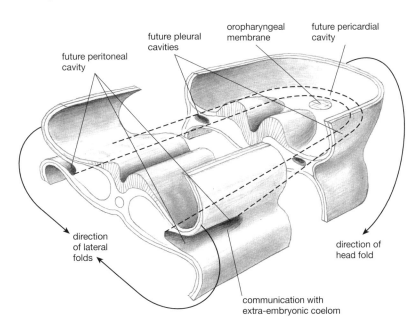

future peritoneal cavity

future pleural cavities

oropharyngeal membrane

future pericardial cavity

direction of lateral folds

direction of head fold

communication with extra-embryonic coelom

4.1 The rostral end of a trilaminar embryo with the roof of the amniotic cavity and the ventral part of the yolk sac cut away, showing the horseshoe-shaped intra-embryonic coelom.

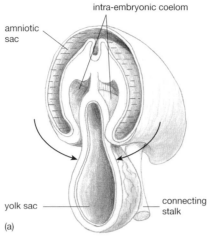

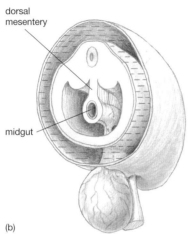

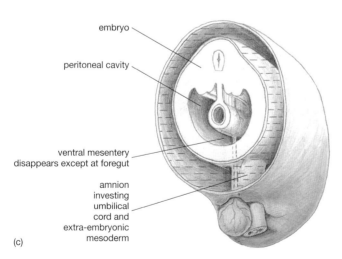

4.2 The developing caudal end of an embryo, to illustrate the lateral foldings that lead to closure of the gut and abdominal cavity (a and b), and the formation of an umbilical cord (b and c).

Transverse folding of the embryo is associated with reduction of the yolk sac to a narrow vitelline duct (**4.2b,c**). The vitelline duct, together with the left umbilical vein, the right and left umbilical arteries and the allantois, which are developing within the extra-embryonic mesoderm of the connecting stalk, all become surrounded by the lining of the amniotic cavity to form the **umbilical cord.**

At the rostral end of the peritoneal cavity the septum transversum (in which the liver develops) forms a ventral mesentery linking the foregut to the anterior body wall. The ventral part of this mesentery will form the **falciform ligament** (in the caudal free edge of which runs the left umbilical vein; see p. 159) and the dorsal part will form the **lesser omentum**, which lies between the stomach and the liver.

The pericardial and peritoneal cavities become separated from the pleural cavities by the formation of two separate partitions, the pleuropericardial and the pleuroperitoneal membranes. The **pleuropericardial membranes** appear rostrolaterally around the common cardinal veins and elongate with the expansion of the lung buds (**4.3**). Subsequent fusion of the membranes on each side with mesoderm ventral to the oesophagus leads to formation of the mediastinum and fibrous pericardium, and separates the pericardial from the pleural cavities.

The **pleuroperitoneal membranes** appear on the caudolateral wall of each pericardio-peritoneal canal and again, with the expansion of the lungs and pleural cavities, these membranes become extended medially and ventrally and are invaded by primitive muscle cells (**4.4**). The two membranes fuse with the dorsal mesentery of the oesophagus and with the septum transversum, forming part of the primitive diaphragm, which separates the pleural cavities from the peritoneal cavity.

The definitive **diaphragm** therefore develops from four structures (**4.5**): (1) the septum transversum, which fuses dorsally with both the mesoderm ventral to the oesophagus and the pleuroperitoneal membranes, forms the central tendon and much of the muscular domes; (2) the pleuroperitoneal membranes which, although they comprise a large part of the primitive diaphragm, form only a small part of the intermediate portion of the fully developed diaphragm; (3) the dorsal mesentery of the oesophagus forms the medial part (the crura) of the diaphragm; (4) the body wall, which is invaded by the expanding lungs and pleural cavity thereby creating an outer layer (the thoracic cage) and an inner layer which contributes to the peripheral parts of the diaphragm. Further expansion of the lungs and pleural cavities into the body wall leads to the formation of the **costo-diaphragmatic recesses.**

The motor and sensory innervation by the phrenic nerve (C3, 4, 5) of the diaphragm, and of the pleura and peritoneum which cover it, reflects the earlier part of its formation opposite cervical somites. Only a small amount of sensory innervation at the periphery is derived from body wall (intercostal) nerves.

aspect of the heart and septum transversum to the primitive pelvic cavity. At this stage the developing peritoneal cavity is still continuous with the pericardial cavity via the pericardio-peritoneal canals at either side of the septum transversum.

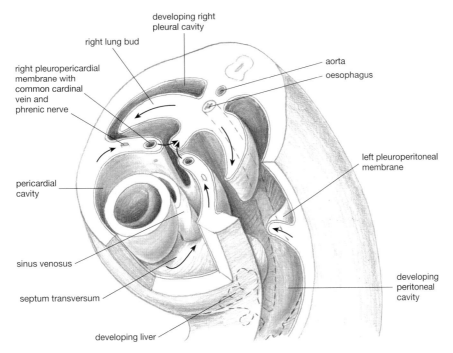

developing right
pleural cavity

right lung bud

right pleuropericardial
membrane with
common cardinal
vein and
phrenic nerve

aorta

oesophagus

pericardial
cavity

left pleuroperitoneal
membrane

sinus venosus

septum transversum

developing
peritoneal
cavity

developing liver

4.3 Separation of the pericardial cavity
from the forming pleural cavities. Note that
the developing lungs invaginate the right
and left pericardio-peritoneal canals.

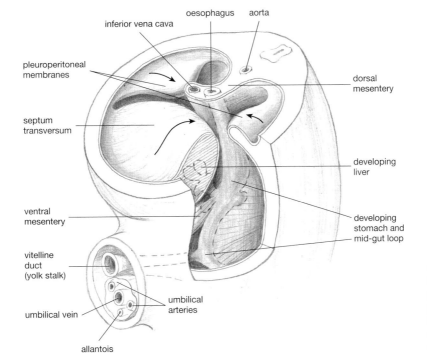

oesophagus   aorta

inferior vena cava

pleuroperitoneal
membranes

dorsal
mesentery

septum
transversum

developing
liver

ventral
mesentery

developing
stomach and
mid-gut loop

vitelline
duct
(yolk stalk)

umbilical
arteries

umbilical vein

allantois

4.4 Separation of the pleural cavities from
the peritoneal cavities by the developing
pleuroperitoneal membranes which, with
growth of the septum transversum, fuse with
mesenchyme around the oesophagus and
elements of the body wall to form the
diaphragm.

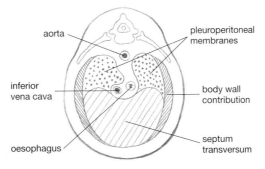

aorta

pleuroperitoneal
membranes

inferior
vena cava

body wall
contribution

oesophagus

septum
transversum

4.5 Developmental components of the
diaphragm.

CHAPTER 5

Thorax: introduction

One of the adaptations that occurred during the time when the primate with an upright stance was evolving, was the development of a shorter and broader chest. The chest wall provides protection for the heart and lungs and forms a closed, but mobile, cage, the volume of which can be increased to cause expansion of the elastic lungs. It is formed of mobile struts—the ribs and their cartilages—which, on either side, articulate posteriorly with the thoracic spinal column and anteriorly with the sternum.

The narrow thoracic 'inlet' between the neck and thorax is closed by a fibrous sheet attached around the trachea, oesophagus, vessels, and nerves to the transverse process of the C7 vertebra and also to the inner margin of the small first rib. The much larger lower margin of the rib cage gives attachment to the muscular diaphragm, which separates the thoracic and abdominal organs.

The diaphragm forms a musculotendinous dome within the rib cage which, by its rhythmic contraction and relaxation, moves down and up to increase and decrease the volume of the chest and hence the lungs. Muscles that move the rib cage, in particular the intercostal muscles between the ribs, influence the volume of the thorax.

The synchronized movements of respiration are co-ordinated by respiratory centres in the brainstem and the associated peripheral nerves, which control the intake and output of air. The air is moved via a series of tubes—the airway—to the alveoli of the lungs, where gaseous exchange with the blood takes place. Within the lungs, both the airway and alveoli are embedded in a very elastic connective tissue which permits expansion during inspiration and expels air by elastic recoil during exhalation.

Separating the surface of each lung from the chest wall is a cavity containing a small volume of serous fluid, the pleural space, which permits friction-free movement between the lung and chest wall. Each pleural space is a closed fibrous sac lined by a serous membrane, into which the lungs are invaginated.

The muscular heart, which pumps blood around the body, is situated in the centre of the chest (mediastinum) between the two lungs. Functionally, the heart has right and left sides: the right atrium and right ventricle which receive deoxygenated blood from the body and pump it through the lungs for gaseous exchange; and the left atrium and left ventricle which receive oxygenated blood from the lungs and pump it around the remainder of the body.

Healthy heart muscle contracts and relaxes regularly from birth to death. Its co-ordinated, rhythmic contractions depend on special conducting tissue, which is itself regulated by the autonomic nervous system. Forward propulsion of the blood also requires competent heart valves to prevent any back flow of blood when the heart muscle contracts. Like the lungs, the heart is surrounded by a serous cavity which permits friction-free movement of the heart. Around this is a fibrous sac (fibrous pericardium). This prevents acute overdistension of the heart chambers. It is sealed around the great outflow vessels (aorta and pulmonary trunk), around the superior vena cava and pulmonary veins, and below to the central tendon of the diaphragm through which enters the inferior vena cava.

Within the mediastinum, the great arteries and veins pass to and from the heart through the thoracic inlet to the head and upper limbs and through the diaphragm to the abdomen and lower limbs. Lymph from the abdomen and lower limbs is carried by the thoracic duct which, on entering the thorax through the diaphragm, passes upward through the posterior mediastinum and leaves via the thoracic inlet to drain into veins in the root of the neck.

The sympathetic trunk and the vagus nerves pass through the thorax, providing autonomic innervation to thoracic organs before continuing into the abdomen. Both the trachea and oesophagus enter the thorax through the inlet. The trachea divides to carry air to and from the two lungs; the oesophagus passes from the inlet downward through the posterior mediastinum and then through the diaphragm to the stomach.

The chest therefore contains two of the organs (heart and lungs) vital to life. Any disturbance of its anatomy, whether by congenital malformation, disease, or trauma, may well have life-threatening functional consequences.

Thoracic cage

Thoracic cage

The bony thoracic cage supports and protects the thoracic viscera and its movements are integral to breathing. The skeleton consists of the thoracic part of the vertebral column, 12 pairs of ribs and costal cartilages, the manubrium and body of the sternum, and the joints between them. Together these enclose the barrel-shaped thoracic cavity. The cavity is much narrower at the top (the inlet) than at the bottom. At the inlet, the thoracic cavity is separated from the neck by the suprapleural membrane. At its base, the thoracic cavity is separated from the abdominal cavity by the diaphragm. This has a domed structure which extends upward within the thoracic cage, so that a substantial part of the upper abdominal organs (liver, stomach, spleen, kidneys) also lie within, and are protected by, the bony thoracic cage.

Development

The development of thoracic vertebrae is discussed with that of the spine (Vol. 1, p. 198). The ribs develop from costal elements of the sclerotome, and the sternum from six sternebrae, the middle four of which fuse together to form the body of the sternum, leaving the manubrium above and the xiphoid process below as separate entities. Occasionally not all the segments fuse, leaving a hole in the body of the sternum. The diaphragm develops mostly from the septum transversum with contributions from adjacent mesoderm (p. 33).

Thoracic cage

The thoracic cage (**5.1.1**) comprises the thoracic spine, 12 pairs of ribs and costal cartilages, and the sternum.

Thoracic spine

The **thoracic spine** comprises 12 thoracic vertebrae with the following characteristics (**5.1.2**):
- 'Heart'-shaped **bodies**, which increase progressively in size from above downward. Each is slightly deeper behind than in front, to produce the normal thoracic

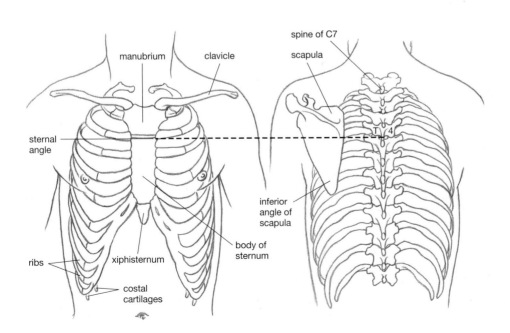

5.1.1 Skeleton of thoracic cage viewed from in front and behind.

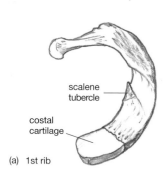

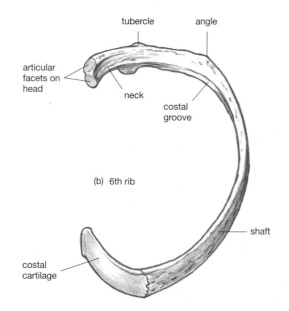

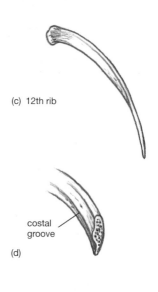

(a) 1st rib

(b) 6th rib

(c) 12th rib

(d)

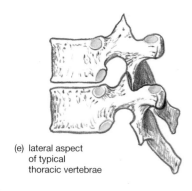

(e) lateral aspect of typical thoracic vertebrae

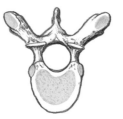

(f) superior aspect of typical thoracic vertebra

5.1.2 (a) First rib; (b) sixth rib. (c) twelfth rib; (d) rib in cross-section; (e) partly articulated thoracic vertebrae, ribs articulate with facets on the transverse processes; (f) superior aspect of thoracic vertebra; note the orientation of the articular facets.

curvature. On each side of the vertebral body are **costal facets** with which the ribs of both sides articulate. The heads of most ribs articulate with the junction of two vertebrae, so there are superior and inferior facets on the lateral aspects of most thoracic vertebrae. However, the first rib articulates only with T1, so it has a complete upper facet (and a lower hemifacet for the second rib); the tenth, eleventh and twelfth vertebrae have only single facets for their corresponding ribs.

- Short, strong **transverse processes** with an anterior-facing facet for articulation with the tubercle of a rib. These facets are absent on T11 and T12.
- Superior and inferior **articular processes** for the adjacent vertebra, orientated in a near-vertical plane. The facet on the lower end of T12 is, however, like an interlocking lumbar facet (Vol. 1, p. 203).
- Long **spinous processes** directed downward.
- **Pedicles** and **laminae**, which together form the **neural arches** that create the posterior wall of the relatively small, round vertebral canal that encloses the thoracic spinal cord.

The vertebrae are connected by:

- **Intervertebral discs** (see Vol. 1, p. 204). These secondary cartilaginous joints limit the movement between individual vertebrae, but permit sufficient movement to allow some rotation of the whole thoracic spine.
- **Synovial joints** between the **articular processes**. The almost vertical orientation of these processes makes **rotation** the principal movement of the thoracic region; some lateral flexion is also possible (flexion would restrict breathing).

Ribs and rib cage

Each **rib** (**5.1.2b**) has a head, neck and body. The **head** of most ribs has two hemifacets for articulation with

the corresponding vertebra and the one above. The ridged area between the hemifacets is attached to the intervertebral disc. The narrowed **neck** connects the head with the **body** of the rib. At the junction is the **tubercle** which articulates with the transverse process of its own vertebra and also has a roughened area for ligamentous attachment to the transverse process. Shortly in front of the tubercle the body of the rib is bent at the **angle**. The inner aspect of the inferior part of most ribs is hollowed to form a **costal groove** in which the intercostal nerve and vessels are located.

At its sternal end, each rib is continuous with a **costal cartilage**. All the ribs slope down from the spine to their costal cartilage, but the costal cartilages slope upward toward their attachment (direct or indirect) to the sternum.

The 12 ribs can be divided into three groups:

- The upper seven ribs (1–7) articulate with the sternum (**vertebrosternal ribs**). (Ribs 1–7 are sometimes called 'true' ribs and ribs 8–12 as 'false' ribs, but this is nonsensical.)
  - The **first rib** (**5.1.2a**) is atypical. It has a wide, horizontally flattened body which slopes downward from its articulation with the first thoracic vertebra. Its costal cartilage (which may ossify late in life) forms a primary cartilaginous joint with the manubrium. The first rib therefore moves little during breathing; rather, it acts as a fixed point of suspension in relation to which the other ribs move. On its postero-lateral surface is a roughened area to which scalenus medius is attached; more anteriorly the **scalene tubercle** is attached to scalenus anterior (Vol. 1, p. 207; both muscles actively prevent depression of the first rib). Behind the scalene tubercle is a smooth groove in which lie the subclavian artery and the T1 root of

the brachial plexus as it passes out of the thorax to enter the axilla.

– **Ribs 2–7** articulate anteriorly with either the manubriosternal junction (rib 2) or the body of the sternum (ribs 3–7) via synovial joints. Posteriorly the heads of these ribs articulate via synovial joints with the body of the numerically corresponding vertebra, and the vertebra above. The capsules of the costovertebral joints are reinforced by fan-shaped ligaments which radiate from the head of each rib to the adjacent vertebral bodies and intervertebral disc (**5.1.3**). Within the joint a ligament attaches the head of the rib to the intervertebral disc. The tubercles of these upper ribs have rounded facets which make synovial joints with saucer-shaped sockets on the transverse processes of the corresponding vertebra (**5.1.2e, f**). The capsules of these joints are reinforced by costotransverse ligaments, which attach the rib to the corresponding transverse process and the transverse process above.

• The **vertebrocostal ribs** (ribs 8–10) are attached to costal cartilages which do not articulate with the sternum but are attached to the costal cartilage of the rib above. Posteriorly, the articulations of the heads of the rib with the vertebral bodies are similar to those of the vertebrosternal ribs, but the facets of the costotransverse joints of these lower ribs are flat, allowing gliding movements to take place (see p. 49).

• The **eleventh** and **twelfth ribs** (**5.1.2c**) are called 'floating ribs' because they are not directly attached anteriorly to the thoracic cage. Posteriorly, ribs 10, 11 and 12 articulate with one vertebra only.

Sternum

The anterior midline of the thoracic cage is formed by the **sternum**. This comprises an upper **manubrium**, a **body**, and a **xiphoid process** (**5.1.4**). The manubrium has a concave upper border which forms the **suprasternal notch**; its lateral border articulates with the clavicle (sternoclavicular joint), with the first costal cartilage, and with the upper facet on the second costal cartilage. The manubriosternal joint between the lower border of the manubrium and the body of the sternum is a secondary cartilaginous joint which may ossify in later life; it forms the **sternal angle**.

The body of the sternum articulates laterally with the second to seventh costal cartilages. The xiphoid process is a cartilaginous flange attached to the lower end of the body of the sternum between the apex of the costal margins; it, too, may ossify in later life.

Thoracic inlet and diaphragm

The **thoracic inlet** is the rather narrow space bounded by the vertebral body of T1, the first ribs and the manubrium. To the right and left of the midline it is closed by **suprapleural membranes** (p. 48). These fibrous sheets are attached to the transverse processes of C7 vertebra and the inner aspects of the first ribs and spread over the apex of the pleura and lung on each side. They separate structures in the neck from those in the **thoracic cavity**.

Inferiorly, the thoracic cavity is separated from the abdominal cavity by the **diaphragm**, a domed fibromuscular sheet which is attached to the xiphoid

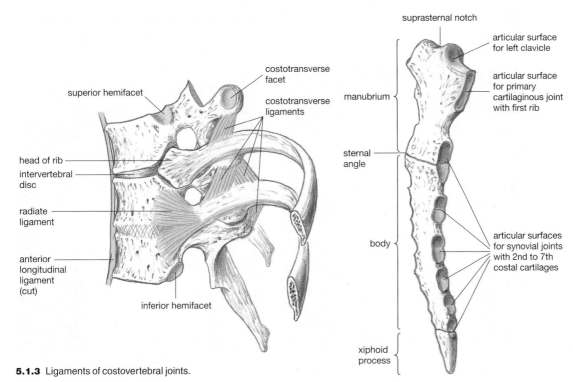

5.1.3 Ligaments of costovertebral joints.

5.1.4 Sternum, lateral aspect, showing sternal angle and facets for ribs.

process anteriorly, the costal margins of each side and the upper lumbar vertebrae, posteriorly (p. 48).

Living anatomy

Inspection

Examine a colleague's thorax from anterior (**5.1.5**) and posterior (**5.1.6**) aspects. It should be almost symmetrical, with symmetrical ventilatory movements. Some chests are long and narrow, others short and broad. On the back, the spine should be examined for any excess anterior curvature (kyphosis) or lateral curvature (scoliosis).

In clinical practice it is important to analyse the character of ventilatory movements (normal, shallow, exaggerated). Indrawing of intercostal spaces on inspiration can indicate obstruction of the airway. In both men and women the breasts should be checked to detect any abnormality (p. 200). The extent of chest hair may indicate disordered androgen secretion.

Palpation

Feel the manubrium and body of the sternum and locate the xiphoid process. Place the tip of your finger just above the centre of the suprasternal notch and press gently backward to feel the trachea. It should lie centrally behind the notch. Gross disease of either the thorax or the root of the neck can displace the trachea from the midline.

Identify the **sternal angle** at the junction of the manubrium and the body of the sternum. The costal cartilages which articulate on either side of the sternal angle belong to the second rib. This is an important landmark from which the numbered ribs and intercostal spaces can be identified. There is relatively little subcutaneous tissue in this region, and even in fat persons it is possible to palpate the sternal angle and thus to locate the second ribs. The first rib is difficult to feel because its sternal end is covered by the medial end of the clavicle. Starting from the second rib, identify all the lower ribs.

Turn to the back and feel down the midline of the neck to locate the prominent spine of the lowest cervical (**C7**) vertebra. Then palpate and count down the spines of all the thoracic vertebrae, noting the degree of curvature of the thoracic vertebral column. Ask your subject to bend forward and touch his/her toes; you will see that no significant flexion takes place in the thoracic region, but is almost entirely confined to the cervical and lumbar parts of the spine.

Qu. 1A *Which is the lowest rib to articulate with the sternum?*

Qu. 1B *Which is the lowest rib to form part of the costal margin?*

Feel the **intercostal spaces** between the ribs anteriorly. In men the nipple usually lies over the fourth intercostal space; in woman, because of the mammary tissue, the position of the nipple is more variable.

To feel the **apex beat** of the heart, place the palm of your hand over the left fifth intercostal space. If you cannot feel the beat, ask the person to lean forward and to the left while sitting, and then feel again. Note the relationship of the apex beat to a line drawn vertically down from the mid point of the clavicle (the **mid-clavicular line**). The apex beat of a normal-sized heart will be at or medial to this line.

Percussion

This is used to provide a rough guide to parts of the chest which are resonant (i.e. filled with air) and those which are filled with solid tissue (such as the heart;

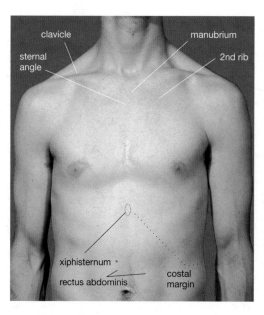

5.1.5 Major landmarks on front of thorax.

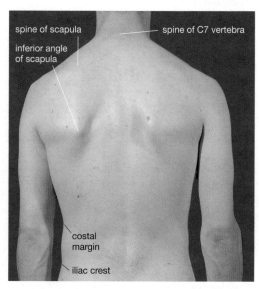

5.1.6 Major landmarks on back of thorax.

or abnormal lung tissue). To percuss a chest (**5.1.7**) lay the palm of your left hand flat on the upper part of the right side of the chest with the middle finger pressed firmly along a rib. Now, keeping your right wrist loose, tap firmly on the middle phalanx of your left middle finger with the tip of your right middle finger (which should be bent). This requires practice but, in this area, you should be able to produce a resonant sound. If you now move your left hand to one of the lower right ribs (i.e. to a rib overlying the liver) and repeat the percussion, the sound you elicit should be duller in tone than before, and you will know you are beginning to percuss successfully. Mark the upper border of the liver dullness.

Percuss down the left anterior chest wall then percuss across the chest from right to left side. There should be a distinct difference in the sound and feel of the vibrations which you generate as you move from areas overlying the lungs to those over the heart. Try to define the right and left borders of the heart, and mark them on your colleague. As you percuss over the lowermost part of the thoracic wall on the left there is no liver dullness, but an area of resonance caused by the gas bubble in the stomach, which may be found just below the apex beat of the heart. Ask your subject to sit up or to lie prone, and repeat the percussion on the back, moving progressively down the left and right sides of the chest.

Compare your findings with an antero-posterior diagram (**5.1.8**) and a radiograph (**5.1.9**) of the chest taken during full inspiration. On full expiration, both domes of the diaphragm rise to about the level of the fifth intercostal space on the anterior chest wall; the heart lies above the central tendon and left dome, the liver lies beneath the right dome, and the stomach beneath the left dome. The lungs fill most of the rest of the thoracic cavity.

Auscultation

Place a stethoscope over the trachea and listen for breath sounds made by movement of air in the trachea and bronchi. Next, place it over the chest immediately below the clavicle and compare the sound made by air entering and leaving the peripheral parts of the lung. Ask your subject to cross his/her arms across the chest and bend slightly forwards, and listen to breath sounds over the posterior chest wall. Note that they decrease in intensity long before the twelfth rib is reached, because the diaphragm and abdominal contents bulge upward into the bony thoracic cage.

Listening to the heart sounds is another important part of routine clinical examination. Place the stethoscope over the apex beat and listen to the 'lub' and 'dup' sounds produced by sudden closure of the heart valves.

Imaging the thoracic cage

Examine 'straight' radiographs of a normal chest (**5.1.9**). 'Straight' radiographs of the chest are usually taken postero-anteriorly (P-A), the subject facing the radiographic film with shoulders protracted to give a wide exposure of the lung fields, and at the height of a deep inspiration. The orientation of chest radiographs will be stated only if it differs from this.

Examine the **bony features**: the bodies of the scapulae should be largely excluded from the lung fields because the shoulders are deliberately protracted during radiography, but the ribs and the borders of the scapulae should be distinct. The vertebral bodies should be faintly visible through the central shadow caused by the structures in the centre of the chest (the mediastinum).

Examine the **soft tissues**: Look for breast shadows (see **5.4.7**). Shadows of the sternomastoid and pectoralis major muscles should also be visible. In the midline locate the radiolucent image of the trachea; this should pass slightly to the right, down to the third or fourth thoracic vertebrae. Compare the appearance of the periphery of the lung fields with the areas where

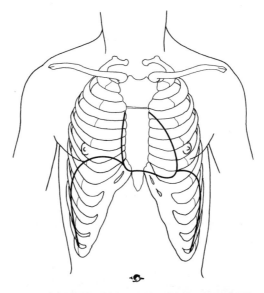

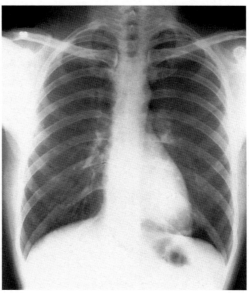

5.1.7 Technique of percussion.

5.1.8 Outline of surface projections of heart and diaphragm in relation to the thoracic cage.

5.1.9 Radiograph (postero-anterior) of thorax in full inspiration.

the lungs join the central mediastinum (roots of the lungs). The radiopaque shadows most prominent in the roots are caused by the presence of blood vessels and sometimes by abnormally enlarged lymph nodes (they are *not* caused by the air-filled bronchi). Examine the apices of the lung fields and the medial, costal, and diaphragmatic boundaries, which should be sharply defined. Note in particular the sharp angle between the chest wall and the diaphragm, often referred to as the costophrenic angle.

Now examine a lateral radiograph of the chest (**5.1.10**). Define the individual thoracic vertebrae, the ribs sloping downward from their vertebral attachments, and the sternum. Identify the domes of the diaphragm, and the shadow created by the heart.

CT and MRI provide sectional images that are now used widely to investigate the organs within the chest and the chest wall (see Thorax, Section 6).

Questions and answers

Qu. 1A *Which is the lowest rib to articulate with the sternum?*

Answer Either the sixth or seventh rib.

Qu. 1B *Which is the lowest rib to form part of the costal margin?*

Answer The tenth rib.

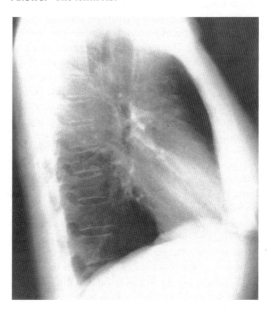

5.1.10 Lateral radiograph of thorax in full inspiration.

The ventilatory pump

The ventilatory pump

For respiration to be effective, oxygen must enter the lungs and carbon dioxide must be removed. The ventilatory pump is responsible for this continuous gas exchange. During development, the right and left lungs invaginate closed serous sacs (pleural sacs) which therefore clothe the outer surfaces of the lungs and line the inner surfaces of the thoracic cage and diaphragm. The two layers of pleura are closely apposed, thus ensuring that when the inspiratory muscles—principally the diaphragm and external intercostal muscles—contract, both the thoracic cavity and the lungs expand and air is drawn in through the upper respiratory tract. Exhalation does not necessarily require any muscular effort as elastic recoil of the lungs and chest wall can suffice to reduce the volume of the cavity when the inspiratory muscles relax. However, the depth of ventilation varies considerably, from quiet breathing at rest to the enormous respiratory movements associated with violent physical exercise. Forcible inhalation and exhalation both involve the actions of additional muscles.

Living anatomy

Movements in quiet and forced breathing

With a colleague sitting, assess the expansion of the upper and lower parts of the chest, and the movements of the abdominal wall during quiet and forced breathing. A simple way to do this is to place your two hands flat on either side of the chest with the thumbs touching in the midline and then ask the subject to breathe in. A better estimate can be gained with a tape-measure placed around the chest at the level of the armpits and then at the level of the xiphoid process; record the minimum and maximum circumference of the chest in quiet and forced breathing.

	Quiet		Forced	
	Min	Max	Min	Max
Upper				
Lower				

The thorax is said to have antero-posterior, transverse, and vertical 'diameters'. The antero-posterior diameter extends from the spine to the manubrium; the transverse, between the most lateral parts of the ribs; and the vertical, from the suprapleural membrane to the diaphragm.

Assess by inspection, and if possible by using calipers, changes in the transverse and antero-posterior diameters of the chest during *quiet* and *forced* breathing. Place your hands, or the calipers, first on either side of the chest, then on the front and back of the chest. Note differences in the movement of the upper chest (vertebrosternal ribs) and lower chest (vertebrocostal ribs).

Qu. 2A *What differences do you observe (a) between the upper and lower parts of the chest; (b) between quiet and forced breathing?*

Qu. 2B *How could you assess changes in the vertical diameter of the chest cavity?*

Exhale deeply and note the contraction of abdominal wall muscles as exhalation proceeds. Cough forcibly, and note the contraction of latissimus dorsi (a broad muscle on the lower back; Vol. 1. p. 55).

The thoracic cavity

The **thoracic cavity** is enclosed primarily by the bony thoracic cage and the diaphragm. Of these, the diaphragm and the ribs are the most mobile.

Because the lungs expand when the chest cavity is expanded, ventilation of the lungs is produced by contraction of the diaphragm, and by movements of the ribs generated by the intercostal and other muscles attached to them. These movements increase the vertical, antero-posterior and transverse diameters of the chest during inspiration and reduce them during expiration (**5.2.1**).

The diaphragm separates the thoracic cavity from the abdominal cavity. It closes the thoracic 'outlet' (bounded by L1 vertebra, the costal margin, and the

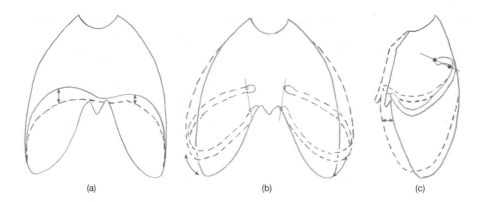

5.2.1 Ventilatory movements of (a) the diaphragm; (b) lower ribs; (c) upper ribs. Red lines mark the axes of movement.

(a)   (b)   (c)

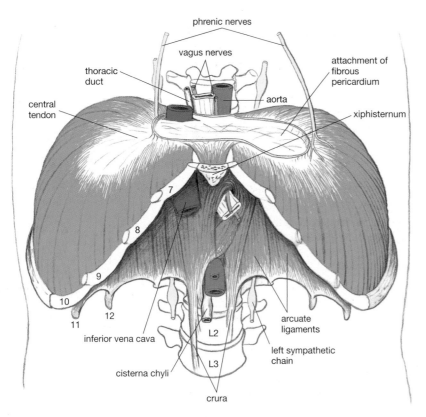

5.2.2 The diaphragm and structures passing through it.

xiphoid process). The thoracic inlet—bounded by the T1 vertebra, the first ribs and the manubrium—is essentially immobile. It is closed laterally by the thick, arching suprapleural membranes attached between the inner aspect of the first rib and the medially placed trachea, oesophagus, and great vessels.

Muscles producing ventilatory movements

The diaphragm

The **diaphragm** (5.2.2) is the principal muscle of inspiration. It forms a dome composed of muscle peripherally and fibrous tissue centrally. It arises by muscular fibres from the inner surfaces of the xiphoid process, the ribs and costal cartilages that make up the

costal margin, and by muscular **crura** which arise from the sides of the bodies of upper lumbar vertebrae (L1, 2, 3—*right*; L1, 2—*left*) and from fibrous bands which link the body of L1 vertebra to its transverse process, and the transverse process to the twelfth rib.

From this origin, muscular right and left **domes** of the diaphragm arch sharply upward within the bony thoracic cage and beneath the bases of their respective lungs. The thoracic surface of the domes is covered by pleura and, at the periphery, each dome make an acute angle with the thoracic wall enclosing the (potential) **costodiaphragmatic recess** of the pleural space. Centrally, the domes are continuous with the trefoil-shaped fibrous **central tendon**, to the upper surface of which the fibrous pericardium is attached.

The diaphragm is supplied by the right and left **phrenic nerves** (p. 84) which originate from C3, 4, 5 spinal nerves in the neck. They descend through the thorax on either side of the mediastinum, pierce the diaphragm, and supply it from its undersurface. They also supply sensory fibres to its pleural and peritoneal coverings, reinforced peripherally by fibres from the lower six intercostal nerves. Small phrenic arteries from the aorta provide its blood supply.

Contraction of the diaphragm causes descent of the domes of the diaphragm and thereby increases the vertical diameter of the thorax and inflates the lungs. This can be visualized by comparing the number of intercostal spaces visible in radiographs of the thorax in full inspiration (5.2.3a) and full expiration (5.2.3b).

In addition, because the lower ribs move like 'buckethandles' (5.2.1b), contraction of the diaphragm also causes the lower chest to increase in transverse (but not antero-posterior) diameter on forced inspiration. This action is dependent on there being a very narrow angle between the diaphragm and chest wall (costodiaphragmatic recess). If the recess is grossly expanded by pathology, contraction of the diaphragm can even pull the lower ribs inward.

The diaphragm itself cannot be seen in plain radiographs of the chest, but the position of the right dome can be inferred from the upper limit of the opacity caused by the liver, and the left from the lower margin of the heart in its pericardium. If, however, gas collects beneath the diaphragm due, for example, to a

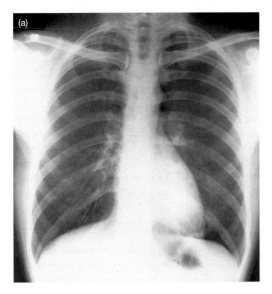

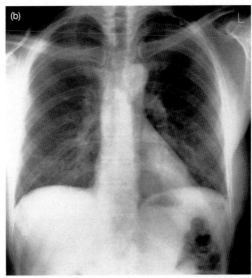

5.2.3 Radiographs of the chest (a) in full inspiration; (b) in full expiration. See Qu. 2C.

perforation of the gut, its dome can be seen clearly (**5.2.4**).

Structures that pass between the thorax and abdomen must either pierce or pass behind the diaphragm. The largest of these are:

- the **inferior vena cava**, which passes through the central tendon at the level of the xiphisternum (body of T8 vertebra) just to the right of the midline.
- the **oesophagus** and **vagus nerves**, which pierce the crura to the left of the midline, at the level of T10.
- the **aorta** and **thoracic duct**, which pass behind the crura in the midline at the level of T12.

Intercostal spaces and muscles

Between the ribs are intercostal spaces (**5.2.5**) which are occupied by the external and internal intercostal muscles (inside the rib cage is an incomplete muscle layer, the innermost intercostals). The intercostal muscles are supplied segmentally by intercostal branches of the thoracic spinal nerves (see **5.2.8**).

The intercostal spaces slope downward and forward between the ribs, but horizontally or upward between the costal cartilages.

External intercostal muscle fibres pass downward and forward from one rib to the upper border of the rib below. Between the costal cartilages, where the orientation of the intercostal space changes, muscle fibres are replaced by a fibrous membrane. Their principal action is in inspiration. The first rib moves very little. The lower vertebrosternal (2–6/7) ribs and vertebrocostal ribs (7/8–10) move differently:

- The anterior part of **vertebrosternal ribs** is pulled upward and forward. They move about an axis which passes between the centres of the head of a rib and of the curved articular facet that its tubercle makes with the transverse process (**5.2.1c**). The principal effect is a small increase in the **antero-posterior diameter** of the upper part of the chest (a movement likened to that of the handle of an old-fashioned village pump). The body of the sternum moves upward

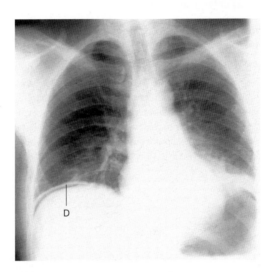

5.2.4 Gas under the diaphragm.

and forward a little, necessitating slight flexion of the manubriosternal joint. The small associated increase in the transverse diameter requires some bending of the costal cartilages.

- The **vertebrocostal ribs** have a plane synovial joint between the rib tubercle and the vertebral transverse process, which permits a gliding movement. These ribs are therefore pulled upward and outward in a bucket-handle-like movement, the axis of which passes between the costal and vertebral articulations. The effect is to increase the **transverse diameter** of the thorax.

As already noted, although it is situated *within* the chest, contraction of the diaphragm pulls the lower ribs upward and *outward* in this 'bucket-handle' movement.

Internal intercostal muscles (**5.2.5**) pass upward and forward from one rib to the rib above; they are replaced posteriorly by a fibrous membrane. They contract strongly during forced expiration.

The intercostal muscles have little action on the floating ribs. However, quadratus lumborum (see **6.3.1**),

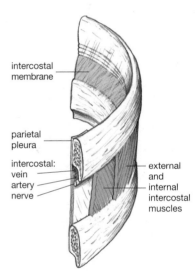

5.2.5 An intercostal space.

Hiccoughs

Hiccoughs are the result of involuntary contractions of the diaphragm. Occasionally, they are caused by irritation of the phrenic nerves by disease, either in the neck or in the thorax.

which arises from the iliolumbar ligament and adjacent iliac crest, inserts into the lower border of the twelfth rib and stabilizes it during respiration.

Qu. 2C *Examine radiographs of a thorax taken in full inspiration (5.2.3a) and full expiration (5.2.3b). What are the major differences?*

Examine also a lateral radiograph of the chest (see 5.1.10), noting again the orientation of the intercostal spaces and the soft-tissue features.

Quiet ventilation

When breathing quietly at rest, the chest wall moves relatively little (though more in women than men). The principal inspiratory muscle expanding the chest and thus the lungs is the diaphragm. As its domes move downward, the elastic anterior abdominal wall expands. Exhalation is almost entirely passive; the diaphragm relaxes and the elastic recoil of the lungs expels air through the airway.

Increased ventilation; accessory muscles of respiration

When ventilation is increased (e.g. during mild exercise) the diaphragm and intercostal muscles contract more forcibly in inspiration, and expiratory muscles (abdominal wall muscles, internal intercostals) act to enhance expiration. However, to prevent a too rapid recoil of elastic lung tissue, some inspiratory muscle activity extends into the expiratory phase to slow the expiratory flow of air.

As ventilation is increased still further, accessory muscles of respiration, such as the scalene muscles of the neck (Vol. 1, p. 207) and limb muscles attached to the chest wall are recruited to assist the increased inspiratory activity. The post-inspiratory activity of the inspiratory muscles is reduced and the activity of expiratory muscles increased.

Take a very deep breath in and note that you automatically **extend** your **thoracic spine** a little. This helps expand the vertical diameter. Now breathe out forcibly and notice that, as expiration proceeds the **anterior abdominal wall muscles** contract more and more strongly to increase intra-abdominal pressure, forcing the relaxed diaphragm upward and reducing the vertical diameter of the chest. Scalene muscles of the neck contract rhythmically to fix and elevate the first rib somewhat.

If even more force is required (often because of lung disease) the **pectoral muscles, serratus anterior,** and **latissimus dorsi** (Vol. 1, p. 61) can all assist in forced inspiration, particularly if the upper limb is fixed by leaning on the arms. In addition, muscles that maintain the patency of, or dilate the upper airway—nostrils, mouth, pharynx and larynx—are also recruited, so that airway diameter is maximal.

Blood supply and lymphatic drainage of the chest wall

The **arterial supply** to the chest wall (5.2.6) is provided by **intercostal arteries** which run along the upper part of each intercostal space.

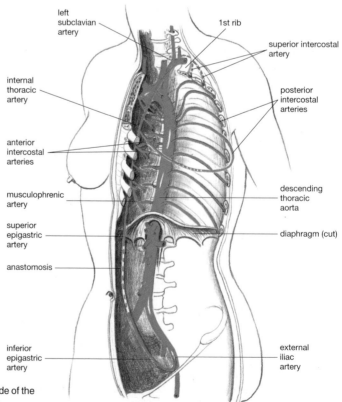

5.2.6 Arterial supply to the left side of the thorax; note sites of anastomoses.

Most **posterior intercostal arteries** are branches of the **descending thoracic aorta**, which runs downward from T4 vertebra in the midline of the posterior aspect of the chest cavity. The posterior intercostal arteries to the upper two intercostal spaces are branches of the superior intercostal artery (the descending branch of the costocervical trunk of the subclavian artery, which arches over the suprapleural membrane to the neck of the first rib). Because the descending aorta lies to the left of the midline for most of its course, the posterior intercostal arteries supplying the right side of the chest have to cross the bodies of the vertebrae.

Small **anterior intercostal arteries** arise from the **internal thoracic artery**, a branch of the subclavian artery which runs downward deep to the costal cartilages. At the lower end of the sternum the internal thoracic artery pierces the diaphragm and divides into the musculophrenic artery and the superior epigastric artery.

The musculophrenic artery runs laterally close to the costal margin, giving branches to the intercostal spaces and diaphragm; the superior epigastric artery continues downward into the sheath of rectus abdominis where it anastomoses with the inferior epigastric branch of the external iliac artery. This link between subclavian and external iliac arteries can provide a collateral pathway if the descending aorta is locally constricted by a defect in development (coarctation of the aorta; see p. 71).

The **venous drainage** of the major part of the chest wall (5.2.7) passes into the **azygos vein** (azygos = unpaired). This vein originates from small lumbar veins on the posterior abdominal wall, passes through the aortic opening of the diaphragm on the right side of the aorta and soon increases in size due to the entry of posterior intercostal veins from the right side of the chest. Intercostal veins draining the left side of the chest wall drain into superior or inferior hemiazygos veins, which cross the midline behind the aorta to join the azygos vein. The azygos vein lies on the right side of the thoracic vertebral bodies and arches forward over the root of the right lung to end in the superior vena cava.

Veins from the first intercostal space usually drain upward into either the brachiocephalic or the vertebral vein; those from the second and third intercostal spaces form a superior intercostal vein which drains into the azygos vein on the right, and into the left brachiocephalic vein on the left. Small veins accompany anterior intercostal branches of the internal thoracic artery. These unite to form an **internal thoracic vein** which drains into the brachiocephalic vein.

The **lymphatic drainage** of the chest wall follows the blood vessels to nodes along the internal thoracic artery, in the posterior mediastinum, and around the diaphragm. Lymph from the medial part of the breast also drains into internal thoracic nodes (p. 201).

Innervation of the chest wall

Somatic innervation of the thoracic cage is derived from the 12 thoracic spinal nerves. The anterior primary rami of the thoracic spinal nerves form the **intercostal nerves** (5.2.8). These supply:
- the intercostal muscles
- the overlying skin
- the underlying parietal pleura.

The posterior primary rami of these nerves supply a strip of skin down the central part of the back and the spinal extensor muscles, some of which have attachment to the ribs (Vol. 1, p. 207).

Each intercostal nerve has a main branch which runs in the costal groove beneath the rib of the corresponding number, and a collateral branch which runs on the upper border of the rib beneath. At the mid-axillary line, each main nerve gives a lateral

---

'Shingles'

Infection of sensory neurons with the chickenpox virus (herpes zoster) can cause strips of pain and skin vesicles **(shingles)** which outline the area of distribution of an intercostal nerve. In contrast, because adjacent nerves supply overlapping segmental strips of skin, loss of one intercostal nerve root does not result in significant sensory loss.

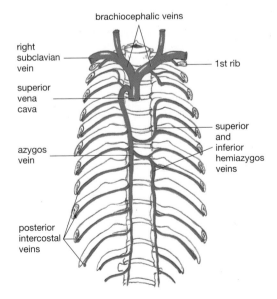

5.2.7 Venous drainage of the posterior thoracic wall (anterior intercostal and internal thoracic veins not shown).

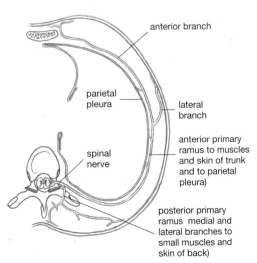

5.2.8 Segmental distribution of an upper thoracic spinal nerve.

branch to the overlying skin; the anterior terminal branch also supplies skin.

The lower six intercostal nerves pass into the anterior abdominal wall where they supply the skin, muscles and parietal peritoneum.

Qu. 2D *Dislocation of the neck at or about the level of C4 is usually fatal. However, a 'whiplash' injury that damages the cervical cord below the level of C4 is compatible with life. Why is this?*

Qu. 2E *If you wished to inject local anaesthetic to relieve the pain of a fractured rib, how would you direct the needle of the syringe, and what structures would the needle traverse?*

Imaging

Examine CTs and MRIs that provide sectional views of the chest wall and diaphragm (see **5.6.1a–f**). On both sets of images, identify as many as possible of the structures you have been studying.

Questions and answers

Qu. 2A *What differences do you observe (a) between the upper and lower parts of the chest; (b) between quiet and forced breathing?*

Answer (a) During *deep* inspiration, in the upper chest, the antero-posterior and transverse diameters both increase a little as the upper ends of the vertebrosternal ribs are raised; in the lower chest, only the transverse diameter increases to any extent. These changes are accompanied by a marked increase in the vertical diameter, caused by the contraction of the diaphragm. The heart and pericardium become more longitudinally elongated as the diaphragm descends. (b) During *quiet* breathing the chest moves much less (especially in males), and respiratory movements are largely accomplished by the diaphragm, which increases the vertical diameter and the transverse diameter of the base of the chest.

Qu. 2B *How could you assess changes in the vertical diameter of the chest cavity?*

Answer On a P-A radiograph of the thorax, measure the distance between comparable points of the diaphragm and clavicle in (a) full inspiration and (b) full expiration.

Qu. 2C *Examine the radiographs of a thorax taken in full inspiration (5.2.3a) and full expiration (5.2.3b). What are the major differences?*

Answer If you count down the ribs you will see that, in full inspiration, the diaphragm descends so that its domes are no higher than the sixth intercostal space anteriorly; in full expiration the domes move up to the fourth intercostal space. Also, the heart lies more horizontally in full expiration.

Qu. 2D *Dislocation of the neck at or about the level of C4 is usually fatal. However, a 'whiplash' injury that damages the cervical cord below the level of C4 is compatible with life. Why is this?*

Answer Injury to the spinal cord below C4 is compatible with life because it leaves the nerve supply to the diaphragm (phrenic nerve C3, 4, 5) largely intact. Paralysis of the diaphragm and intercostal muscles by an injury at C4 would require continuous artificial ventilation to keep the subject alive.

Qu. 2E *If you wished to inject local anaesthetic to relieve the pain of a fractured rib, how would you direct the needle of the syringe, and what structures would the needle traverse?*

Answer The nerve, with its accompanying vessels, lies in the subcostal groove under cover of the lower border of the rib, and deep to both the external and internal intercostal muscles. The needle should therefore be directed upward immediately under the rib. The needle would traverse skin, superficial fascia and both intercostal muscles.

Pleura, lungs, and airways

Pleura, lungs, and airways

The pleural cavity on each side of the chest is a closed serous sac. Each lung invaginates and virtually fills the sac, leaving only a potential space. This creates a parietal layer of pleura which lines the thoracic cage, the dome of the diaphragm, and the mediastinum; and a visceral layer which covers the lung. The trachea and main bronchi are situated in the mediastinum, but their smaller divisions lie within the lung, together with their terminal sacs (alveoli) for gas exchange and the vessels and nerves of the lung. The connective tissue of the lung is very elastic. It is stretched when the lungs expand during inspiration, and its elastic recoil is the most important factor in unforced expiry of air.

Development of the lungs and lower airways

At the beginning of the fourth week of fetal life a median groove (**laryngotracheal groove**) and corresponding ridge (**laryngotracheal ridge**) appear on the ventral wall of the foregut (5.3.1). Immediately caudal to this an outgrowth (diverticulum) develops. The endoderm of this **laryngotracheal diverticulum** forms the epithelial surface and glandular tissue of the larynx, trachea, and entire lower respiratory tract; the surrounding mesoderm forms connective tissue, smooth muscle, and cartilage of the lower respiratory tract. As the groove deepens and expands, it becomes separated from the part of the foregut that will form the oesophagus by **tracheo-oesophageal folds**, which grow in laterally to unite in the midline. The lumen of the developing oesophagus becomes occluded and must later recanalize.

At the caudal end of the laryngotracheal diverticulum, a **lung bud** develops. This divides into right and left **bronchial buds**. These buds grow laterally to bulge into the medial walls of the **pericardio-peritoneal canals** (the primitive pleural cavities). The buds divide repeatedly to generate the primordia of the **primary bronchi**, and continue to divide until **terminal**

Developmental abnormalities of the tracheo-oesophageal junction

If the partitioning between the developing trachea and oesophagus is incomplete, then different types of tracheo-oesophageal connections (fistulas) can occur. Such anomalies often coexist with oesophageal malformation and, after birth, swallowed material can enter the airway, causing coughing, choking, and respiratory distress.

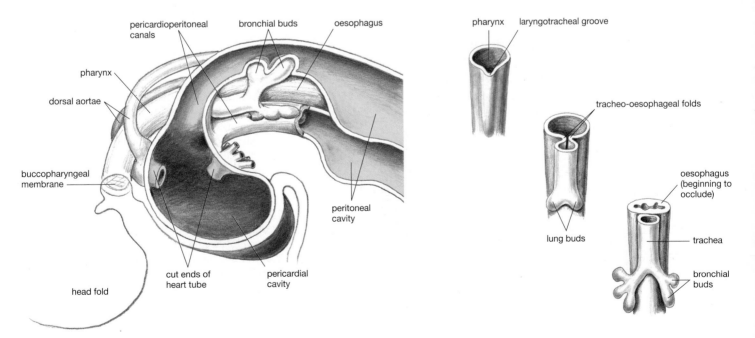

5.3.1 Development of the trachea and of the lung buds as they start to invaginate the pericardio-peritoneal canals.

bronchi are formed. By about the sixth month, **terminal sacs** (the primitive alveoli), surrounded by pulmonary vasculature derived from the sixth aortic arch, begin to appear. This is, therefore, the very earliest stage at which premature infants can survive. Further development of the terminal sacs produces both the thin squamous epithelial cells optimal for gaseous exchange (this thinning continues to 8 years of age) and type 2 epithelial cells which secrete **surfactant** when stimulated by corticosteroids. The surfactant lipoprotein is necessary to lower the surface tension within the alveoli and thus enable them to remain patent when air enters the lungs. If surfactant secretion is inadequate, as is common in premature infants, then the alveoli cannot expand with air, and respiratory distress results.

Qu. 3A *Why are the lung fields in a radiograph of the chest of an infant more radiopaque than those in an adult?*

General disposition of thoracic organs

Removal of the anterior chest wall (**5.3.2**) reveals the **right** and **left lungs**, surrounded by their **pleural membranes** and **pleural sacs**, lying on either side of the centrally placed **mediastinum**. In the mediastinum, the largest organ is the **heart**. This is covered by a **fibrous pericardial sac** attached below to the central tendon of the diaphragm and above to the outer coverings of the great vessels. Anteriorly the lungs and pleura overlap the pericardium almost to the midline, so that only a small part of the pericardium is visible in a concavity of the anterior border of the left lung, the **cardiac notch**.

For description, the mediastinum is divided into:

- An **anterior mediastinum**, situated in front of the heart between its fibrous pericardium and the chest wall. In this space are the internal thoracic arteries, veins and lymphatics, the thymus gland, and some fatty connective tissue (see also p. 83).
- A **middle mediastinum**, which contains the heart, pericardium and proximal parts of the great vessels, the roots of the lungs (where the airways, vessels, and nerves enter/leave the lungs), and the phrenic nerves which run downward from the neck on either side to the heart to reach the diaphragm.
- A **posterior mediastinum**, situated behind the heart. It contains the oesophagus, the descending aorta and the main vessels supplying and draining the thoracic cavity, the main lymphatic channel (the thoracic duct), and elements of the autonomic nervous system.
- A **superior mediastinum**, situated above the heart and behind the manubrium (i.e. above the level of the sternal angle). It contains structures passing between the thorax and either the neck or the upper limb (the trachea and oesophagus, the arch of the aorta and its branches, the superior vena cava and its tributaries, lymphatics, phrenic and vagus nerves, and the sympathetic trunk).

Re-examine a plain P-A chest radiograph (see **5.1.9**) and remind yourself of the bony landmarks, the diaphragm and the costophrenic angle that it makes with the rib margin.

The trachea should be visible as a translucent shadow passing down from the midline of the suprasternal notch to lie slightly to the right of the midline at T3/4. The hilus of each lung appears relatively radiopaque when compared with the lung fields, because it contains large blood vessels.

The lung fields should appear progressively more radiolucent from the hilus to the periphery. Air in the alveoli does not obstruct the passage of X-rays, so bronchi cannot be seen because there is air inside and outside them, and their walls are thin The radiopaque areas normal peripheral lung fields is caused mainly by small blood vessels. Any accumulation of fluid or solid tissue will appear as abnormal radiopaque areas.

Pleural sacs

The left and right **pleural sacs** are two closed sacs which, during the course of development, are invaginated by the growing lungs. The **pleural membranes** which line the sacs therefore form an outer layer of **parietal pleura** which lines the chest wall, suprapleural membrane, diaphragm and mediastinum, and a layer of **visceral pleura** which clothes the lung. Between these two layers, which are closely apposed, is the **pleural cavity**, which contains only a thin film of fluid. Each pleural cavity thus acts like a large bursa that

5.3.2 Lungs and great vessels after removal of the anterior thoracic wall.

Labels: trachea · oesophagus · internal jugular vein · common carotid artery · subclavian vein · subclavian artery · remnant of thymus gland · transverse fissure · upper lobe · upper lobe · oblique fissure · oblique fissure · middle lobe · cardiac notch · lower lobe · lower lobe · fibrous pericardium reduces friction between the lungs and chest wall as they move relative to one another during respiration.

The lower border of the pleural sac, where the parietal pleura lining the chest wall becomes continuous with that covering the diaphragm, lies at the level of the eighth rib in the mid-clavicular line, the tenth rib at the mid-axillary line, and the twelfth rib posteriorly. Inside the pleural sac the lower border of the lung with its covering of visceral pleura lies about two ribs higher, at the level of the sixth, eighth and tenth ribs, respectively. Between these two levels the pleural cavity provides a potential space—the **costodiaphragmatic recess**—which allows for downward expansion of the lungs (5.3.3).

Around the root of the lung, the parietal and visceral pleural membranes become continuous; inferiorly, this reflection is 'loose' to allow for downward movement of the lung roots during inspiration (the 'pulmonary ligament'; see **5.3.6**).

During inspiration the thoracic cavity is expanded along with its attached lining of parietal pleura. The pressure within the pleural cavity, which is lower than that of the atmosphere, then becomes even more subatmospheric. This forces the lungs to expand and the air pressure in the alveoli and respiratory tract falls, drawing air through the nose and mouth into the conducting airways and alveoli (provided that the airway is patent).

Elastic recoil of the lungs and chest wall makes unforced expiration largely a passive process (see p. 50). The cartilage in the walls of the trachea and bronchi prevents the tendency of these parts of the airway to collapse.

It is important to remember that, normally, the pleural cavities are only potential spaces. They can, however, be distended by accumulations of air, blood, pus, or other material. Any accumulation causes a reduction in the volume available to the lung and hence less efficient lung function.

Radiograph **5.3.4** shows a large accumulation of fluid in the right pleural cavity. The horizontal upper border (fluid level) of the opacity shows that there is also air in the pleural cavity.

If the pleural membranes become inflamed, the lung may become attached to the chest wall by fibrous **adhesions**.

Innervation of the pleura

The parietal pleura is innervated by somatic nerves and is very sensitive (intercostal nerves supply pleura lining the chest wall; phrenic nerves supply that lining the diaphragm and mediastinum). Any inflammation of the parietal pleura (pleurisy) causes pain which is intensified by breathing movements.

By contrast, the visceral pleura, which forms the outer connective tissue of the lung, is not supplied by somatic sensory nerves. Cutting the visceral pleura therefore does not cause pain. Only autonomic fibres reach the visceral pleura.

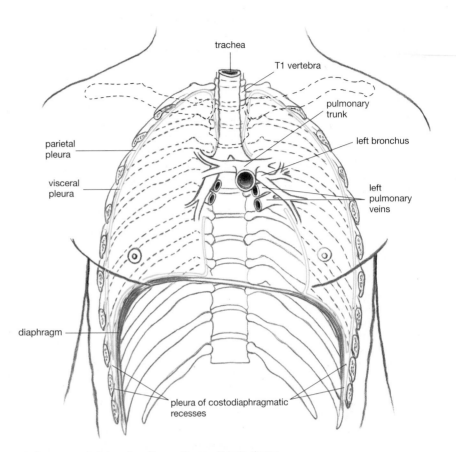

5.3.3 Arrangement of pleural cavities and lungs within the thorax.

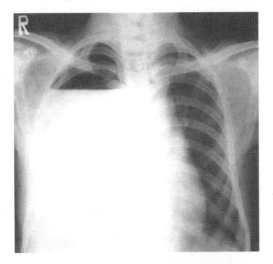

5.3.4 Radiograph of thorax, showing fluid in the right pleural cavity. Note the fluid level visible in this upright patient.

The lungs

The lungs are light, spongy and pink in the young. In the adult, however, the tissue becomes a mottled black as a result of inhaling carbon particles (especially in tobacco smokers).

Each lung is pyramidal in shape (5.3.6). The highest point of its rounded **apex** lies about 2 cm *above* the

Pneumothorax

Anything that penetrates the surface of a lung (for example, a fractured rib which tears the parietal and visceral pleura and penetrates lung tissue, or spontaneous rupture of surface alveoli) can allow air to escape into the pleural cavity. If this happens, the lung will collapse by reason of its natural elasticity, and a **pneumothorax** results. Air may also escape through the parietal pleura into the tissues of the chest wall, causing them to swell, sometimes quite dramatically.

If the puncture of the lung acts as a valve, continued inspirations will cause air to accumulate in the pleural cavity, which then will be under increasing positive pressure, a condition called 'tension pneumothorax'. This condition can rapidly become fatal because the heart and great vessels are pushed to the opposite side of the chest, compressing the opposite lung and seriously embarrassing the action of the heart.

Radiograph **5.3.5** is of a patient with a pneumothorax. Note the difference in appearance between lung tissue and air in the left pleural cavity.

mid point of the clavicle. It therefore appears to project above the inlet to the thorax, but this is largely an illusion created by the oblique orientation of the first rib. The lung apex is separated from structures of the neck by the dome-shaped suprapleural membrane attached to the inner aspect of the first rib and the transverse process of C7.

The **costal surface** of the lung, which is very rounded posteriorly where it occupies the paravertebral recess, follows the contour of the thoracic cage to end in a sharp anterior border. On the right this lies vertically close to the midline between the second and the sixth costal cartilages; on the left the cardiac notch diverges from the midline at the fourth costal cartilage; on both sides the sharp inferior border is related to the eighth rib in the mid-axillary line, and the tenth rib posteriorly.

The concave **diaphragmatic surface** of the lung arches up and over the dome of the diaphragm. Its **mediastinal surface** fits around the heart and vessels lying medial to it. In lungs fixed in preservative fluid, indentations of the lung surface caused by these structures can often be clearly seen. At about the mid point of the mediastinal surface, the **root** or **hilum** (5.3.6) is found, which transmits structures entering or leaving the lung.

Each lung is divided into an **upper lobe** and a **lower** (posterior) **lobe** by an **oblique fissure** which runs along a line starting at the spinous process of T3 and passing downward and forward around the costal surface to the base of the lung at the level of the sixth costal cartilage. The right lung is also divided by a **horizontal fissure** which runs from the oblique fissure in the mid-axillary line, horizontally forward to the level of the fourth costal cartilage; it demarcates the **middle lobe** of the right lung. These fissures may be incomplete and are usually not visible on chest radiographs.

Each lobe is subdivided into a number of **segments**, each of which is supplied by a branch of the bronchial

tree (5.3.7, 5.3.8a, b). Details of most segments are not clinically important, but a few points should be noted:
- The apical segments of the upper lobe of each lung project some 2 cm above the mid point of the clavicle.

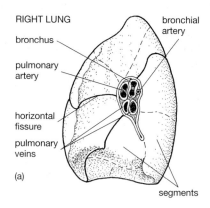
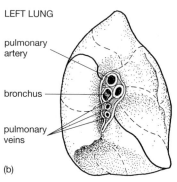

5.3.6 Structures entering and leaving the roots of the lungs.

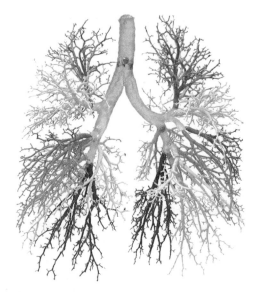

5.3.7 Plastic cast of trachea and bronchopulmonary segments. Segmental bronchi are colour coded. **Upper lobe:** apical, brown; posterior, grey; anterior, pink. **Middle lobe** (R) **or lingula** (L): lateral (R), superior (L), blue; medial (R), inferior (L), red. **Lower lobe:** apical, pale green; medial basal, yellow; anterior basal, light grey; posterior basal, dark grey.

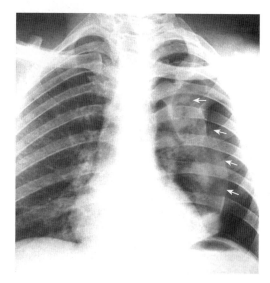

5.3.5 Pneumothorax of the left pleural cavity. The left lung has partially collapsed (its lateral border is *arrowed*) due to the presence of air in the pleural cavity.

- The apical segments of the lower (posterior) lobes lie high up on the posterior chest wall (see **5.3.9**) and tend to collect fluid in recumbent patients because their bronchi pass directly posteriorly (**5.3.8b**).
- The basal segments of the lower lobes form the base of each lung. If small objects are inhaled when standing upright, they tend to pass into the basal segments, particularly on the right, because the right main bronchus is wider and more vertical.

Qu. 3B *Examine radiograph* **5.3.10.** *What abnormality do you see?*

Roots of the lungs (5.3.6)

Structures entering or leaving each lung via the roots, are:
- The **main bronchus**. That on the right divides into a bronchus to the upper lobe which lies above the main artery (eparterial bronchus) and a bronchus to the lower lobe.
- The **pulmonary artery**, which carries deoxygenated blood from the pulmonary trunk to the lungs. Each major artery lies inferior to the main bronchus but, on the right, gives a branch which runs with the bronchus to the upper lobe before entering the lung root.
- **Pulmonary veins** (1–4 in number), which carry oxygenated blood to the left atrium of the heart. These lie anterior and inferior to the bronchi.
- Small **bronchial arteries** that arise from the aorta lie on the posterior aspect of each main bronchus; they supply the bronchial tree with oxygenated blood.
- The **pulmonary plexus** of nerves which surrounds the bronchi. It consists of fibres from the vagus nerve, which cause contraction of bronchial smooth muscle, stimulate secretion from mucous glands, and supply sensory (e.g. stretch, compression) receptors; and sympathetic fibres from T1–T5 ganglia, which supply bronchial muscle, secretory glands, and blood vessels.
- **Lymphatics**, which drain the substance of the lung and enter nodes in the lung root (hilar nodes); these in turn drain to nodes lying around the bifurcation of the trachea (tracheobronchial nodes), which form a mediastinal trunk that returns lymph to the venous system. The hilar nodes are often black because of the inhaled carbon particles the lymphatics have taken up.

Many diseases of the lungs, including tuberculosis and carcinoma, involve the lymph nodes, which can enlarge sufficiently to cause obstruction of a bronchus and collapse of a lobe (see **5.3.10**).

The airways

The great variation that occurs in gas exchange with exercise requires that flow through the airways can be varied considerably.

The airway consists of the nose and mouth, naso- and oro-pharynx, larynx, trachea, bronchi and bron-

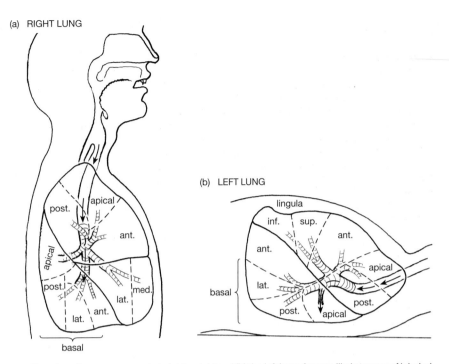

5.3.8 Bronchopulmonary segments in (a) the right and (b) the left lung. Arrows, likely course of inhaled material: (a) in upright position and (b) in recumbent position.

chioles. These convey gas in and out of the bronchioles and alveoli where gas exchange occurs.

Quiet breathing usually occurs through the nose (unless this is obstructed). The nasal air passages are largely enclosed in bone and, apart from flaring of the nostrils, cannot change in diameter. When increased gas exchange is required, the mouth is used as an additional airway. If, however, the nasal mucosa becomes very swollen, with an upper respiratory tract infection, enlarged nasopharyngeal lymphoid tissue (adenoids) or hay fever, even quiet breathing has to occur through the mouth.

The laryngeal airway is enclosed in cartilage and so varies little in diameter, but the extent of opening of the glottis (Vol. 3, Chapter 6, Section 6) varies considerably with the depth of breathing.

The anterior and lateral walls of the **trachea** contain ⌒-shaped cartilaginous plates which hold it open; its posterior wall is smooth muscle which can vary the diameter a little. **Bronchi** also have cartilaginous plates in their wall to maintain patency, but their smooth muscle encircles the bronchi and can therefore constrict them. The cartilage becomes progressively less as bronchi become smaller.

Qu. 3C *What would be the effect on the bronchi of a stress that activates the sympathetic system?*

The entire airway (excluding the mouth, glottis, terminal bronchioles, and alveoli) is lined with ciliated respiratory mucous epithelium. Dust particles become trapped in the mucus and the cilia move the mucus toward the pharynx, where it is swallowed. Cigarette smoke paralyses the cilia, thus allowing secretions to accumulate; these can become infected, hence the 'smoker's cough'.

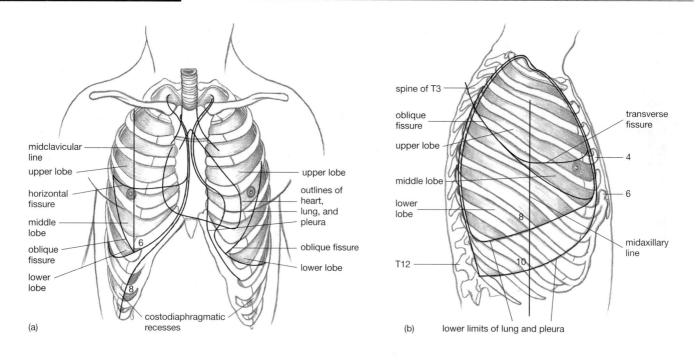

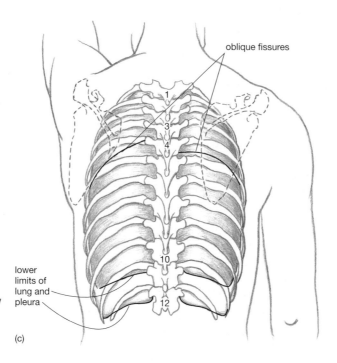

5.3.9 (a) Anterior, (b) lateral, and (c) posterior aspects of the rib cage, to show the relationship to surface landmarks of the heart, the lungs and their fissures, and the lower limits of the pleura.

Coughing is a protective reflex (p. 230) which helps to clear the respiratory tract. Usually it is preceded by a deep inspiration followed by a forced expiration against a glottis (vocal cords) which is initially closed, but then opens rapidly to allow a sudden expulsion of a column of air which can dislodge a foreign body in the airway. The reflex is triggered by irritation of neural receptors in the respiratory mucosa.

Qu. 3D *Which neurotransmitter could be administered to reduce the spasm and improve breathing?*

The **trachea** enters the thorax almost in the midline. It passes downward and backward and divides, at the **carina**, into right and left **main bronchi** behind the sternal angle (about T4 level).

• The **right main bronchus** (see **5.3.3, 5.3.7, 5.3.8**) is shorter, wider, and more vertical than the left in the adult and is therefore more likely to receive inhaled foreign bodies. Immediately before entering the lung it gives off the bronchus to the upper lobe; it continues into the lung then divides into bronchi serving the middle and lower lobes.

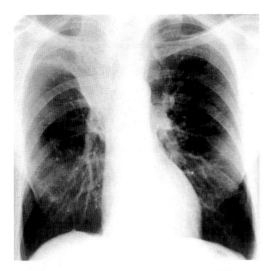

5.3.10 Abnormal radiograph of thorax (see Qu. 3B).

- The **left main bronchus** divides into an upper-lobe bronchus and a lower-lobe bronchus only after entering the lung. The upper-lobe bronchus also supplies the **lingula** (equivalent to the right middle lobe).

  In both lungs, the first segmental bronchus to arise from the posterior aspects of the bronchial tree passes to the apical segment of the lower lobe. Therefore, if a recumbent patient inhales vomit or secretion, this segment of the lung (usually on the right) is commonly affected with pneumonia or abscess formation.

Qu. 3E *How would you position a patient to allow satisfactory natural drainage of this region?*

Broncho-pulmonary segments (see **5.3.7, 5.3.8**) are independent functioning units, each containing a segmental bronchus with its accompanying pulmonary vessels. They can be defined by imaging techniques and, if necessary, excised surgically.

Qu. 3F *Which lobe of which lung is a peanut accidentally inhaled at a party most likely to enter?*

Radiographs **5.3.11** taken on deep inspiration (left) and forced expiration (right), are of a child who has inhaled a small apple core.

Qu. 3G *Explain the radiographic findings.*

Bronchoscopy

The interior of the airway can be examined by using a bronchoscope. **5.3.12a** shows the carina at the bifurcation of the trachea; **5.3.12b** shows the origin of the right main bronchus, and **5.3.12c** shows an obvious deviation from normal. **5.3.12d** shows a tumour of the bronchus which is bleeding slightly.

Living anatomy

Re-examine the expansion of the chest that occurs during inspiration, and percuss again the anterior and posterior chest wall. Listen to the breath sounds, thinking now of the lobes of the lung from which the sounds originate (**5.3.9**). Because of the obliquity of the upper lobe of the lung, breath sounds can readily be listened to by placing a stethoscope on the anterior aspect of the chest above the fourth costal cartilage. If the stethoscope is then placed between the vertebral borders of the scapulae and the vertebrae, sounds originating in the upper parts of the lower (posterior) lobes of the lung will be heard. If your subject raises his right arm over his head and you listen between imaginary lines passing forward from the mid-axillary line along the fourth and sixth costal cartilages, you will hear sounds from the middle lobe of the right lung.

You should now be able to mark on a colleague: (1) the contour of the diaphragm in full expiration and inspiration; (2) the lower limits of the lungs and pleura; (3) the apices of the lungs; (4) the oblique and transverse fissures of the right lung; and (5) the oblique fissure of the left lung.

Imaging

CT imaging (see **5.6.1**) taken with the usual soft-tissue setting reveals the trachea and major bronchi as air-filled spaces, but shows no detail of the lung tissue (**5.3.13a**). However, the method can be manipulated ('lung window setting') to visualize details of lung tissue (**5.3.13b**). MRI imaging is rarely used for the respiratory system as the air gives no signal.

Qu. 3H *Using your knowledge of anatomy, what would be the most likely symptoms or results if the tumour was to press on or invade:*

(a) a bronchus;
(b) a pleural cavity;
(c) the oesophagus;
(d) the pericardium;
(e) the superior vena cava;
(f) the left recurrent laryngeal nerve;
(g) the right phrenic nerve;
(h) the sympathetic chain (stellate ganglion) in the apex of the thorax;
(i) the thoracic duct.

Questions and answers

Qu. 3A *Why are the lung fields in a radiograph of the chest of an infant more radiopaque than those in an adult?*

Answer The lung fields are more dense in an infant because thinning of the alveolar septa continues during the first few years of life. The thinning increases the ratio of air to soft tissue, and hence the ability of X-rays to penetrate the chest.

Qu. 3B *Examine radiograph 5.3.10. What abnormality do you see?*

Answer The upper lung field on the right side is very dense due to collapse of the right upper lobe.

Qu. 3C *What would be the effect on the bronchi of a stress which activates the sympathetic system?*

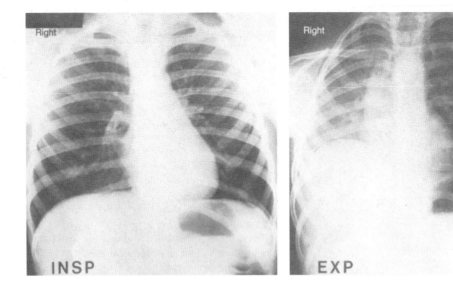

5.3.11 Radiographs taken during forced inspiration (INSP) and expiration (EXP) (see Qu. 3G).

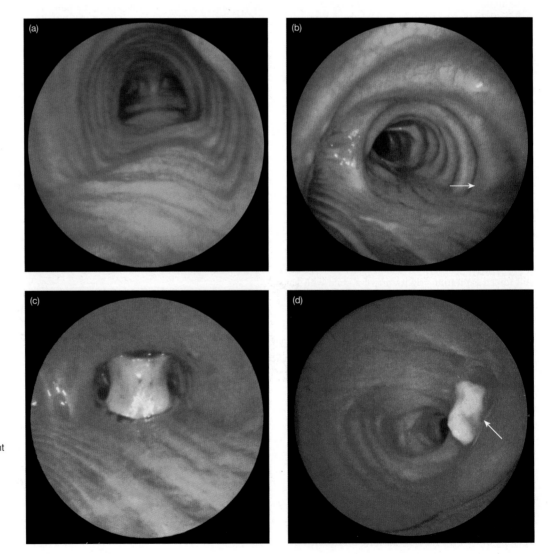

5.3.12 Bronchoscopic views of (a) the trachea and its bifurcation into left and right main bronchus at the carina; (b) the right main bronchus with opening of the right upper lobe bronchus (*arrow*); (c) chicken vertebra lodged in right main bronchus; (d) carcinoma (*arrowed*) in right main bronchus at the opening of upper lobe bronchus.

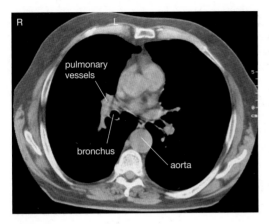

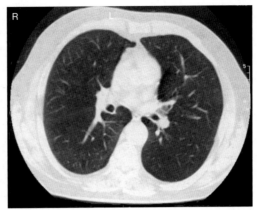

5.3.13 Axial CTs at T5/6 level taken with (a) soft-tissue setting and (b) 'lung window' setting to show details of the lung tissue.

Lung cancer

The most common type of intra-thoracic tumour is that arising in bronchial mucosa. Moreover, in industrial Western countries such tumours are among the most common to be found in men and women. Their incidence rose dramatically in the twentieth century, due largely, it is thought, to exposure to carcinogens in tobacco smoke. Many of the symptoms produced by malignant tumours are due to pressure or invasion by a growing tumour, or by lymph nodes to which the cancer has spread, on other intra-thoracic structures.

Answer Stress stimulates the output of adrenaline and noradrenaline and causes bronchodilatation via β-adrenergic receptors on the smooth muscle fibres. This enables increased ventilation during stress.

Qu. 3D *Which neurotransmitter could be administered to reduce the spasm and improve breathing?*

Answer Adrenaline or noradrenaline, or drugs that mimic their action, would dilate the bronchi.

Qu. 3E *How would you position a patient to allow satisfactory natural drainage of this region?*

Answer Lay the patient prone, with the head turned to one side (the 'recovery position').

Qu. 3F *Which lobe of which lung is a peanut accidentally inhaled at a party most likely to enter?*

Answer The right lung, because of the larger size and vertical orientation of the right main bronchus.

Qu. 3G *Explain the radiographic findings.*

Answer The apple core has entered the left main bronchus. The radiograph taken in inspiration appears normal; on expiration, however, the left lung does not deflate, the right lung deflates excessively, and the mediastinum is pushed to the right. Dilatation of the bronchus allows some air to pass around the obstruction during inspiration, but relaxation of its wall on expiration allows the bronchus to close around the obstruction creating a 'one-way valve'. Respiratory movements on the right are deeper in an attempt to compensate for the lack of gaseous exchange on the left side.

Qu. 3H *Using your knowledge of anatomy, what would be the most likely symptoms or results if either the*

tumour was to press on or invade:

(a) a bronchus;
(b) a pleural cavity;
(c) the oesophagus;
(d) the pericardium;
(e) the superior vena cava;
(f) the left recurrent laryngeal nerve;
(g) the right phrenic nerve;
(h) the sympathetic chain (stellate ganglion) in the apex of the thorax;
(i) the thoracic duct.

Answer

(a) Pressure on a bronchus would lead to collapse of the segment that it supplies.
(b) Invasion of the pleural cavity would cause an effusion of pleural fluid which, if sufficiently large, would embarrass respiration.
(c) Pressure on the oesophagus would cause difficulty with swallowing.
(d) Invasion of the pericardium would lead to a pericardial effusion which, if sufficiently large, would reduce venous return to the heart and therefore cardiac output.
(e) Pressure on the superior vena cava would obstruct venous drainage of the head, neck and arms, which would be swollen.
(f) Pressure on the left recurrent laryngeal nerve would cause a hoarse voice.
(g) Pressure on the left phrenic nerve could lead to paralysis of the left dome of the diaphragm.
(h) Pressure on the sympathetic chain in the apex of the thorax could disrupt the sympathetic supply to the head, causing, on the affected side, a constricted pupil, drooping eyelid, loss of sweating, vasodilatation, and a sunken eye (enophthalmos).
(i) Invasion of the thoracic duct would lead to an effusion of lymph into the pleural cavity (chylothorax).

Heart and great vessels

Heart and great vessels

The heart comprises two muscular pumps arrayed in series, but working synchronously. Each pump consists of two chambers (an atrium and a ventricle) in series. In postnatal life the right heart pumps deoxygenated blood to the lungs; the left heart pumps oxygenated arterial blood at higher pressure around the rest of the body. Valves located between the atria and ventricles and in the right and left outflow tracts prevent back-flow. The potentially autonomous contractions of the atria and ventricles are co-ordinated by specialized conducting tissue, which bridges the fibrous septum which insulates the ventricles from the atria, and are modulated by the autonomic nerves to the heart. Coronary arteries convey oxygenated blood to the heart muscle; its venous blood returns to the right atrium. The heart and its vessels are enclosed within a sac of pericardium, which together form the middle mediastinum.

Development of the heart

See **5.4.1, 5.4.2**.
During the third week of fetal life, mesenchymal cells lying ventral to the pericardial coelom begin to aggregate on either side of the embryo to form two parallel cardiogenic cords. The cords canalize to form endothelial tubes (endocardial tubes) which then fuse to form a single midline **heart tube**. Other mesenchymal cells condense around the heart tube to form potential heart muscle (**myocardium**) and the surrounding visceral **pericardium**. Filling the potential space between the endocardium and myocardium is connective tissue, termed **cardiac jelly**.

With the increasing curvature of the head fold, the heart tube, suspended in the pericardial part of the coelomic cavity (p. 31) by a dorsal mesocardium (which soon breaks down), comes to lie caudal to the oro-pharyngeal membrane, i.e. ventral to the foregut and anterior to the septum transversum.

The heart is anchored at its rostral end by the developing aortic arches, which pass around the foregut to join the dorsal aorta, and at its caudal end by the entry of three systems of paired veins (vitelline, umbilical, and cardinal, draining, respectively, the yolk sac, placenta, and fetal body wall), which pass through the septum transversum into the sinus venosus.

While the single heart tube is forming, a series of dilatations—the **truncus arteriosus, bulbus cordis, ventricle, atrium**, and **sinus venosus**—develop in a rostral to caudal direction. As a result of differential growth, the heart tube attains a U-shape and then becomes S-shaped, with the atrium and sinus venosus lying dorsal to the ventricle, the bulbus cordis, and the truncus arteriosus.

Autonomous contractions of the myocardium start at about the 22nd day and, within a few more days, blood begins to flow from the venous to the arterial end of the heart.

During the fourth week of development, the cardiac jelly forms **atrioventricular endocardial cushions** on the dorsal and ventral walls of the canal leading from the atrium to the ventricle. These cushions fuse in the middle of the canal to create a septum, which divides the atrioventricular canal into right and left channels. The right and left (tricuspid and mitral) atrioventricular **valves** also form from this cushion tissue.

Septation of the atria

At the same time as the endocardial cushions are developing:
- A thin, crescentic **septum primum** grows down from the roof of the atrium towards the cushions and eventually fuses with them (**5.4.2**). The decreasing gap between the septum and the cushions is called the **foramen primum**.
- Before the foramen primum closes, apoptosis (programmed cell death) causes perforations to appear in the dorsal part of the septum primum. These coalesce to form the **foramen secundum**.
- Toward the end of the fifth week a second crescentic partition, the **septum secundum**, grows downward to the right of the septum primum to cover the foramen secundum. The two horns of the crescent make contact with cushion tissue, leaving an oval aperture, the **foramen ovale**, which, with the foramen secundum, remains patent until birth, forming a flap-like valve.

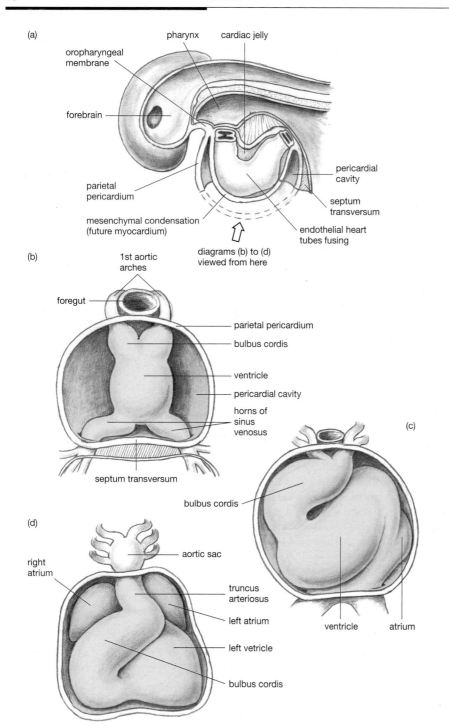

5.4.1 Early development of the heart tube.

pressed against the septum secundum, closing the foramen ovale and forcing blood in the right atrium to flow into the right ventricle for distribution to the lungs.

Septation of the ventricles

At the same time as the interatrial septum is forming, partitioning of the common ventricle commences:
- A thick crescentic fold, the **interventricular septum**, develops largely as a result of the flow of two streams of blood through the atrioventricular channels on either side of the endocardial cushion tissue. The posterior horn of the crescent is continuous with the posterior endocardial cushion, and the anterior horn with the anterior bulbar ridge. It forms the muscular part of the interventricular septum.
- Partitioning of the bulbus cordis and truncus arteriosus divides the aorta from the pulmonary trunk and separates the outflow tracts of the two ventricles. **Bulbar** and **truncal ridges** develop from the cardiac jelly on the right and left walls of the bulbus and truncus, and fuse to form an **aorticopulmonary septum**. The septum is spirally orientated (again due largely to the streams of blood flowing from the developing right and left ventricles). Its lower end fuses posteriorly with the atrioventricular endocardial cushion and anteriorly with the muscular interventricular septum. At the junction of the bulbus cordis and truncus, the aortic and pulmonary valves develop by thinning of endocardial cushions formed by bulges of cardiac jelly.
- Complete separation of the two ventricles (which occurs about the end of the seventh week) is achieved by a downgrowth from the right **endocardial cushion** which fuses with both the muscular interventricular septum and aorticopulmonary septum. This forms the membranous part of the interventricular septum.

Development of the great vessels (5.4.3)

During the fourth and fifth weeks of fetal life, bilateral **branchial arches** form around the foregut, into which paired arterial arches, arising from the truncus arteriosus, grow in a dorsal direction. These **aortic arches** join the bilateral dorsal aortae, which fuse below the level of the eleventh somite to form a single **dorsal aorta**.

Although, theoretically, six pairs of aortic arch arteries are formed, they are not all present at the same time because, as caudal arches form, so some rostral arches degenerate. Between the sixth and eighth week the adult pattern is achieved:
- the first two arch arteries disappear almost entirely;
- the third arch arteries form the **common carotid** arteries proximally and the first part of the **internal carotid** arteries distally (the **external carotid** arteries branch from the third arch arteries later, when the face develops);
- the fourth arch artery on the left side forms most of the **arch of the aorta**, while that on the right becomes

---

Atrial septal defects

The foramen ovale often fails to close completely after birth but, because pressures in the right and left atria are both low, little blood passes through the defect and there are few consequences. Large defects which permit substantial mixing of blood are much less common.

The foramen ovale enables oxygenated blood entering the right atrium (derived from the umbilical circulation), which is at a higher pressure than blood in the left atrium, to flow through the foramen ovale into the left atrium and thence primarily towards the head, rather than into the relatively closed pulmonary circulation.

At birth, however, when the placental circulation ceases and the baby's lungs are filled with air, the pulmonary circulation opens, increasing the pressure in the left atrium. The septum primum then becomes the **brachiocephalic trunk** (by joining proximally with the third arch artery) and the initial part of the **right subclavian** artery.

- the fifth arch arteries do not develop;
- the proximal part of the sixth arch artery on the left side forms the stem of the **left pulmonary artery**. Its distal part forms the **ductus arteriosus** which provides a shunt that carries blood from the pulmonary trunk to the aorta during fetal life when the lung circulation is virtually closed. The proximal part of the sixth arch artery on the right side persists as the stem of the **right pulmonary artery**; its distal part degenerates, as does the dorsal aorta on the right between the third arch and the point of union of the two dorsal aortae. On either side, more distal parts of the pulmonary arteries grow from the sixth arches as the lungs develop.

The vagus (X) nerve supplies the sixth arch. Its recurrent laryngeal branch on the left side hooks around the ductus arteriosus before ascending to reach the larynx. On the right, however, because the fifth and the distal part of the sixth aortic arches do not develop, the right recurrent laryngeal nerve hooks around the more rostrally placed right subclavian artery derived from the fourth arch.

Development of the venous system

The part of the inferior vena cava just beneath the heart is formed from the conjoined right umbilical and right vitelline veins. The superior vena cava is formed from the right common cardinal vein and the adjacent part of the right anterior cardinal vein.

Draining into the superior vena cava are the posterior cardinal vein (arch of the azygos vein), the remainder of the right anterior cardinal vein (right brachiocephalic vein), and a cross-midline anastomosis between the right and left anterior cardinal vein (left brachiocephalic vein).

Both venae cavae open into the smooth-walled part of the right atrium, which is derived by incorporation of the right horn of the sinus venosus. The coronary sinus, formed from the left horn of the sinus venosus, also drains into the right atrium.

The pulmonary veins develop as an outgrowth from the left atrium toward the developing lung buds. The outgrowth divides repeatedly into branches which drain the developing lungs. The first four branches become absorbed, forming the smooth-walled part of the left atrium, so that four pulmonary veins normally open into that chamber.

The heart

The **heart** is a four-chambered muscular pump. In postnatal life it comprises two pumps working in series:

- the **right heart** pump—the **right atrium** and **right ventricle**—receives deoxygenated blood from the body tissues and pumps it into the pulmonary circuit for oxygenation and removal of carbon dioxide;

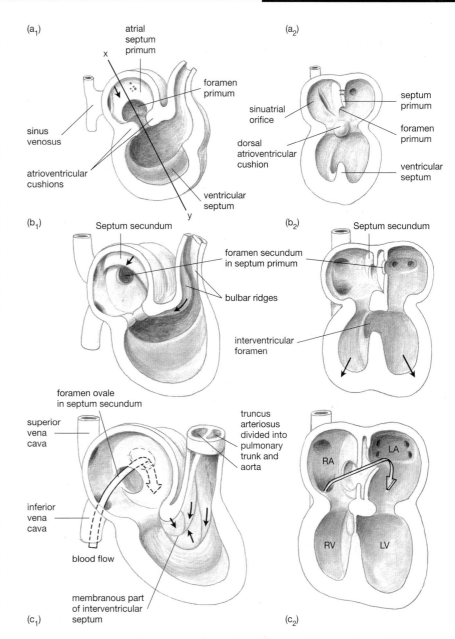

5.4.2 The developing heart bisected sagittally and viewed from the right (a_1, a_2, a_3) and bisected coronally and viewed from the front (b_1, b_2, b_3). The diagrams show the division (septation) of the right and left atria and ventricles and the formation of the aorta and pulmonary artery from the bulbus cordis.

- the **left heart** pump—the **left atrium** and **left ventricle**—receives oxygenated blood from the lungs and pumps it around the rest of the body.

Either of these pumps can fail and the consequences are very different for right and left heart failure. When the right heart fails the systemic veins are not fully emptied, the liver becomes enlarged, and there is peripheral oedema, most noticeable in the (dependent) lower limbs. When the left heart fails, the pulmonary veins are not fully emptied and fluid accumulates in the lungs (pulmonary oedema).

The heart is situated in the middle mediastinum, between the lungs, enclosed in a sac of fibrous and serous **pericardium**. It lies obliquely in the thoracic

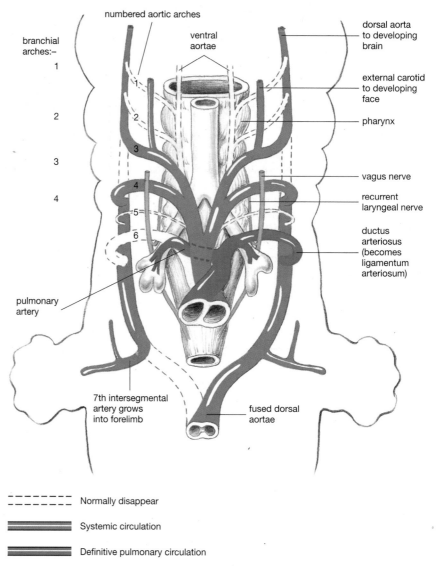

branchial arches:–
1
2
3
4

numbered aortic arches

ventral aortae

dorsal aorta to developing brain

external carotid to developing face

pharynx

vagus nerve

recurrent laryngeal nerve

ductus arteriosus (becomes ligamentum arteriosum)

pulmonary artery

7th intersegmental artery grows into forelimb

fused dorsal aortae

– – – – – –   Normally disappear

▬▬▬▬▬   Systemic circulation

▬▬▬▬▬   Definitive pulmonary circulation

5.4.3 Development of the great arteries.

Ventricular septal defects

The ventricular and aorticopulmonary septa (most commonly the cushion component which forms the membranous part of the interventricular septum) may be incomplete.

Qu. 4A In which direction would blood flow through a defect in the interventricular septum?

If the bulbar ridges fail to divide the aortic and pulmonary channels equally, they are also likely to fail to unite with the endocardial cushions. In the severe abnormality known as **Fallot's tetralogy**:
- the developing pulmonary outflow tract is too narrow;
- the aorta therefore overrides both the right and left ventricles and the interventricular foramen remains patent;
- the narrowing of the pulmonary tract causes hypertrophy of the right ventricle;
- the resultant mixing of venous and arterial blood lowers the oxygen tension of the arterial blood (cyanosis), a condition commonly referred to as 'blue baby'.

Qu. 4B *If the pulmonary artery is very narrow, how could blood reach the lungs after the infant is born? (see below)*

Qu. 4C What other relatively common congenital abnormalities of the heart would you expect from a knowledge of its development?

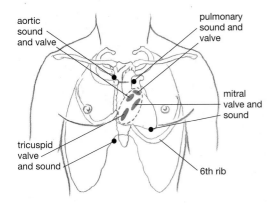

5.4.4 Orientation of heart, surface markings of heart valves, and sites for best auscultation (sound) of mitral (M), tricuspid (T), aortic (A) and pulmonary (P) valves.

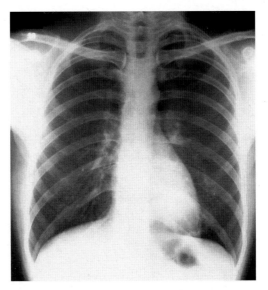

5.4.5 P-A radiograph of normal chest in full inspiration.

Disorders of ductus arteriosus closure

If the ductus fails to close—**patent ductus arteriosus**—blood in the aorta will enter the lower-pressure pulmonary circulation after birth. If a patent ductus is large enough, the turbulence of blood passing between the two great vessels gives rise to remarkably loud sounds which, through a stethoscope, resemble machinery in action.

If, however, the process of closure (which is stimulated by prostaglandins) extends too far into the aorta, a constriction—**coarctation of the aorta**—develops either just before or just after the origin of the left subclavian artery. Such an obstruction is partially by-passed through dilated left internal thoracic and left intercostal arteries (see **5.2.6**). The enlarged intercostals cause localized pressure on the ribs and resorption of bone, which appear as notches on radiographs of adults.

cavity behind the sternum and adjoining parts of the rib cage, one-third to the right of the midline and two-thirds to the left (**5.4.4, 5.4.5**). It has:

- a **base**, which is apposed to the posterior mediastinum;
- a diaphragmatic (inferior) surface, which rests on the central tendon of the diaphragm;
- a convex anterior sterno-costal surface;
- an **apex**, which forms its lower left extremity, situated normally behind the fifth intercostal space at or within the mid-clavicular line.

When viewed from the front, the heart has:

- a right border formed by the right atrium;
- an inferior border formed largely by the right ventricle, with the left ventricle at the apex;
- a left border formed almost entirely by the left ventricle (the auricle of the left atrium overlaps its upper end).

The anterior surface of the heart is formed mostly by the right ventricle; the right atrium and its auricle are to its right, and a part of the left ventricle and the left auricle to its left. The left atrium forms the major part of the base (posteriorly); and the right and left ventricles form the majority of the diaphragmatic surface.

Living anatomy

The most important surface landmark is the **apex beat** of the heart. You can feel it by placing the flat of your hand on the left side of the subject's chest at the level of the nipple or just below (fourth to fifth intercostal space). If you have difficulty feeling the apex beat, get your subject to lean forward and to the left in a sitting position (to move the heart closer to the chest wall), or to step on and off a chair several times (to increase the force of contraction), and then again feel for the apex beat.

Assess the relationship of the apex beat to the mid-clavicular line. The apex beat normally lies inside this line, which in most adults is about 8.5 cm from the sternum.

Percuss the margins of the heart (**5.4.4**), thinking of the chambers involved. Try to define:

- the **right border** of the heart (formed by the right atrium), which extends between the articulations of the third and fifth costal cartilages with the sternum;
- the **left border** of the heart (formed mainly by the left ventricle), which runs obliquely downwards and laterally from the second costal cartilage to the apex;
- the **inferior border** of the heart (formed mainly by the right ventricle), which extends from the junction of the fifth right costal cartilage to the apex.

Draw the shape of the heart on your subject and on a rough outline of the chest which you should now be able to sketch with ease. In a tall thin, person the heart is somewhat more vertical than in someone of a short, stocky build.

Listen to the heart sounds with a stethoscope. At the apex you should be able to hear the normal 'lub' and 'dup' sounds of each cardiac cycle. The 'lub' is caused by closure of the atrioventricular valves as the ventricles contract at the beginning of systole; the somewhat sharper 'dup' sound by closure of the aortic and pulmonary valves as the ventricles relax at the beginning of diastole.

The apex is the best place to listen to sounds from the left atrioventricular (mitral) valve. To hear sounds of the tricuspid valve most clearly, place the stethoscope over the lower left sternal edge. Listen also over the right and left borders of the manubrium, where you will best hear sounds made, respectively, by the aortic and pulmonary valves. The sites at which you can best hear sounds from a valve are *not* the surface markings of the valves (see **5.4.4**), but reflect conduction of the sounds by the skeleton and soft tissues.

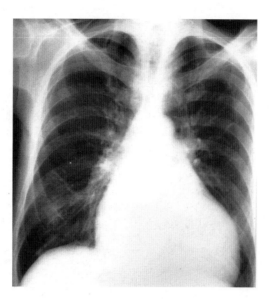

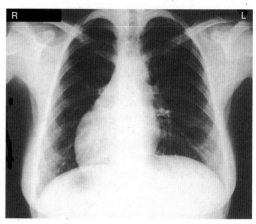

5.4.7 P-A radiograph of thorax. Compare with **5.4.5** and see Qu. **4E**.

5.4.6 P-A radiograph of thorax. Compare with **5.4.5** and see Qu. **4D**.

Examine the size and shape of the heart on both P-A (**5.4.5**) and lateral (**5.1.10**)radiographs. The transverse diameter of a normal heart should not exceed half the total width of the chest.

Qu. 4D *Examine radiograph 5.4.6. What abnormality is visible?*

On a P-A radiograph of a normal chest (see **5.4.5**) trace the right border of the mediastinum, which is comprised, from above downwards, of the right brachiocephalic (innominate) vein, the superior vena cava, the right atrium and a very short segment of the inferior vena cava. Next trace the left border from the arch of the aorta (which, viewed P-A, is known as the aortic knuckle), past the left pulmonary artery, left atrium and left ventricle to the apex.

On a lateral radiograph (see **5.1.10**), the posterior margin of the heart is formed by the left atrium. Immediately behind it lies the oblique sinus of the pericardium and the oesophagus.

Qu. 4E *Examine radiograph 5.4.7. What developmental anomaly is present?*

The 'right heart'

External appearance (5.4.8)

The thin-walled **right atrium** forms the right border of the heart. Its ear-like extension, the **right auricle**, overlaps the base of the pulmonary artery. The right atrium receives deoxygenated blood from:

- the **superior vena cava**, which drains the upper limbs, head, and thoracic walls and enters upper extremity of the right atrium;
- the **inferior vena cava**, which drains the abdomen, pelvis, and lower limbs and enters the atrium immediately after passing through the diaphragm;

- the **coronary sinus**, which, with other small veins, drains the walls of the heart and enters the atrium close to the opening of the inferior vena cava.

The right atrium and right ventricle are separated by the **right atrioventricular sulcus**, which contains the right coronary artery.

The **right ventricle** forms most of the sternocostal and part of the diaphragmatic surface of the heart. It gives rise to the **pulmonary trunk** which passes upward to the left of the aorta for a short distance then divides beneath the arch of the aorta into right and left **pulmonary arteries** which pass to the lungs.

The right ventricle is separated from the left ventricle by an **anterior interventricular sulcus** in which a branch (anterior interventricular) of the left coronary artery passes obliquely down and to the left toward the apex of the heart.

Internal features (5.4.9)

The openings into the right atrium of the superior and inferior venae cavae and of the coronary sinus have no functional valves but there is normally little backflow because emptying of the right atrium occurs primarily as a result of relaxation of the right ventricle (rather than atrial contraction).

The anterior wall of the right atrium and its auricle appear rough because of bundles of muscle (musculi pectinati) in its wall. The bundles converge on a vertical band of muscle fibres, the **crista terminalis** which extends between the opening of the **superior vena cava** and the opening of the **inferior vena cava**. The remainder of the right atrial wall and the interatrial septum is smooth walled. The crista terminalis marks the boundary of the incorporation of the sinus venosus (smooth wall) into the primitive atrium; the sinuatrial node ('pacemaker') is situated at its upper end. The **coronary sinus** (derived from the left horn of the sinus venosus) opens into the atrium just to the left of the inferior vena cava.

The **interatrial septum** is marked by an oval depression, the **fossa ovalis**. A small, oblique channel in the fossa is sometimes present; this is the remains of the fetal interatrial connection, which enables blood to bypass the lungs.

Blood from the right atrium passes into the right ventricle through the **right atrioventricular (a-v) valve (tricuspid valve)** during diastole. When the ventricles contract, the a-v channel is closed by the three **cusps** (leaflets) attached to its anterior, posterior, and septal margins. The free margins of the cusps are tethered by fine strands of connective tissue (**chordae tendineae**) to **papillary muscles** that project from the walls of the ventricle (**5.4.9**). Valve cusps are composed of fibrous tissue covered with endothelium. They are attached to an annular fibrous skeleton derived from a-v cushion tissue (**5.4.10**) which separates the atria from the ventricles and also forms a part of the interventricular septum.

Qu. 4F *What is the function of the chordae tendineae and papillary muscles?*

Qu. 4G *If the tricuspid valve cannot close properly during systole, what will be the effect on the right atrium? If the defect is considerable, what will be the effect on the liver and the lower extremities?*

The **right ventricle** has a thicker wall than the right atrium. Muscle bundles project into its lumen as thick ridges (trabeculae carneae); conical projections form papillary muscles to which the cordae tendineae are attached. One bundle at the base carries a branch of the conducting system (moderator band).

The funnel-shaped outflow tract (**infundibulum**) of the right ventricle is smooth-walled and leads upward and slightly to the left, to become continuous with the pulmonary artery which carries the deoxygenated blood to the lungs. The **pulmonary valve** lies at the junction of the infundibulum and pulmonary artery. It consists of three delicate semilunar concave cusps (**5.4.10**), each with a nodule thickening the mid point of its rim. During systole the cusps are flattened against the wall of the artery. At the end of systole, when the pressure in the ventricle falls, blood in the artery tends to return to the ventricle, filling the cusps and bringing together their free margins to close the valve.

The 'left heart'

External appearance (5.4.8)

The **left atrium** forms most of the posterior surface of the heart, its small auricle curving round the left border on o the anterior surface. It receives the **pulmonary veins** carrying oxygenated blood from the lungs.

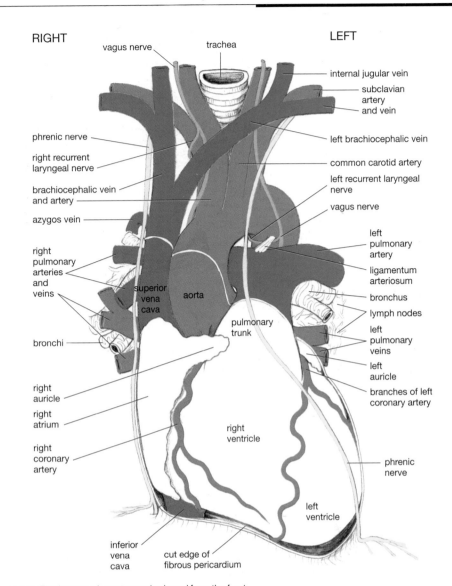

5.4.8 The heart and great vessels viewed from the front.

The thick-walled **left ventricle** forms only the left margin of the sternocostal surface of the heart, but a considerable part of its diaphragmatic surface. The left atrioventricular sulcus contains the trunk (circumflex branch) of the left coronary artery and, on the diaphragmatic surface. The anterior interventricular sulcus contains the anterior interventricular branch of the left coronary artery; the posterior interventricular sulcus contains a branch of (usually) the right coronary artery.

The root of the elastic-walled **ascending aorta** leaves the upper aspect of the left ventricle behind the pulmonary trunk, to emerge on its right side. Three bulges in its wall—the **aortic sinuses**—each accommodate a cusp of the aortic valve during systole. The **right** and **left coronary arteries** originate from two of these sinuses (see **5.4.12**).

Qu. 4H *Why is the muscular wall of the left ventricle normally three times thicker than that of the right ventricle?*

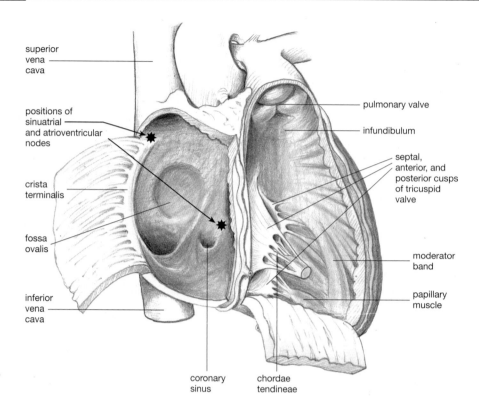

5.4.9 Interior of right atrium and ventricle viewed from the right side.

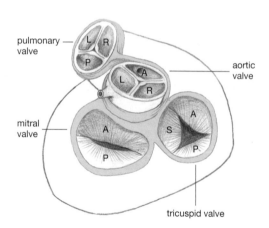

5.4.10 Valves of the heart viewed after removal of the atria and aorta and pulmonary trunks. The fibrous tissue which separates the atria from the ventricles and supports the atrioventricular valves is continuous with that which supports the valves at the roots of the great vessels. A, anterior; P, posterior; R, right; L, left; S, septal.

Internal features (5.4.11)

The **left atrium** is smooth walled apart from its auricle. It receives oxygenated blood from the four **pulmonary veins** which enter its posterior surface. The **left atrioventricular valve** has only two leaflets (anterior and posterior), shaped rather like a bishop's mitre, and is therefore called the **mitral valve**. Its attachment to the fibrous skeleton, chordae tendineae, and papillary muscles is similar to that of the tricuspid valve.

Qu. 4I *If the mitral valve became narrowed (stenosed) what would be the effect on the heart and lungs?*

The interior of the **left ventricle** is larger than that of the right ventricle and is markedly trabeculated, except for a smooth-walled outflow tract leading to the **aortic valve** and the root of the **aorta**.

The aortic valve (5.4.10, 5.4.11) has three semilunar cusps, similar to those of the pulmonary valve. Anatomical texts describe one anterior and two posterior cusps, but in the fetus and in echocardiography, two anterior and one posterior cusp are commonly described.

Immediately above the cusps, the wall of the aorta is dilated to form **aortic sinuses** which each accommodate a valve cusp during systole. During diastole these dilatations cause turbulence in the blood which, by pushing against the cusps, helps ensure proper closure of the valve.

Note that, whereas the a-v valves have the papillary muscles and chordae tendineae to prevent 'blow-back' during ventricular contraction, the outflow valves rely only on the shape of their valves.

Any calcification in a damaged heart valve will appear as a radiopaque shadow at the site of the valve.

Qu. 4J *If a person was found to have a dilated left atrium, what abnormality of a heart valve would you expect? What radiological view of the heart would most easily reveal this condition?*

Blood supply of the heart

The heart, in particular the left ventricle, has a profuse blood supply (5.4.12–5.4.14). The main flow in the arteries occurs during diastole because, during systole, contraction of ventricular muscle compresses the capillaries.

Two main arteries supply the heart. The **right coronary artery** arises from the aortic sinus above the

Heart valve abnormalities

Any heart valve may either be narrowed ('stenosed') or fail to close properly ('incompetent'). The abnormality may be the result of congenital defects, or may develop during adult life for a variety of reasons, among which are rheumatic heart disease and hypertension. Any abnormality reduces the pumping efficiency of the heart, but the effects on the different parts of the circulation will vary depending on which valve is involved.

Stenosis of a valve will cause turbulence of blood flow and an abnormal heart sound ('murmur'). This will occur during diastole if an a-v valve is involved and during systole if a pulmonary or aortic valve is involved. Incompetence of a valve will allow reflux of blood and murmurs from both types of valve will occur during diastole.

A combination of the cardiovascular effects and the heart sounds enables a diagnosis of the valvular problem(s) to be made.

If a valve becomes inefficient (stenosed or incompetent), it can be replaced by an artificial valve or by one from a suitable animal.

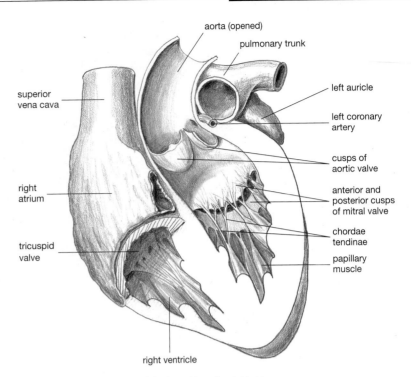

5.4.11 Interior of left atrium and ventricle viewed from the right side.

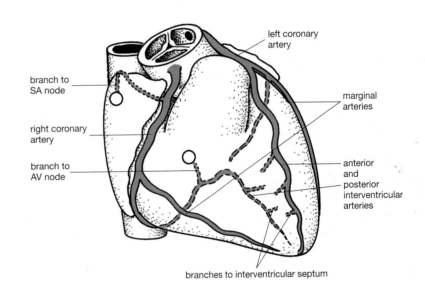

5.4.12 The arterial supply of the heart.

anterior cusp of the aortic valve; the **left coronary artery** from the sinus above the left posterior cusp (for this reason clinicians often refer to a right, a left, and a non-coronary sinus). The arteries pass, respectively, around the right and left atrioventricular sulcus. From this incomplete arterial 'ring' four main branches arise, which all pass towards the apex of the heart.

- the **anterior interventricular artery** (anterior descending branch) is the largest; it arises from the left coronary artery and supplies the left and right ventricles and the septum;
- the **posterior interventricular artery** (posterior descending branch) passes to the apex of the heart along the posterior interventricular sulcus and also supplies both ventricles and the septum. It usually arises from the right coronary artery but may arise from the left ('left dominant' pattern);
- the **right marginal artery** arises from the right coronary artery and passes along the inferior margin of the heart, supplying the right ventricle and apex;
- the **left marginal artery** arises from the left coronary artery and runs along the left border of the heart, supplying the left ventricle.

The part of the left coronary artery which curves around the left border of the heart is often referred to as the 'circumflex branch'; it supplies the left atrium and ventricle and may (40%) supply the sinu-atrial node. Both the sinu-atrial and atrioventricular nodes are more commonly supplied by the right coronary artery.

All the main coronary arteries are effectively end-arteries to the cardiac muscle, although there is some anastomosis between their smaller branches.

The coronary arteries receive some autonomic innervation, primarily from the sympathetic system which can cause both vasodilation (β) and vasoconstrictor (α) effects, but nervous regulation is much less important than local metabolic control.

Veins accompany all the branches of the coronary arteries (5.4.14). Most eventually drain into a large vessel, the **coronary sinus**, which runs in the atrioventricular groove on the posterior aspect of the heart before opening into the right atrium. Some venous blood may also pass directly into the atrial chambers from the walls of the heart via small veins (venae cordis minimae; anterior cardiac veins).

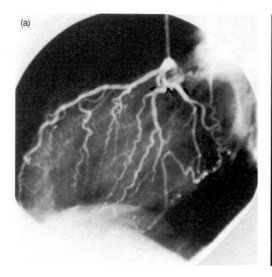
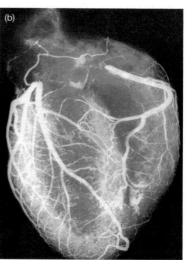

5.4.13 (a) Left coronary angiogram; left posterior oblique view. (b) Angiogram showing both coronary arteries; postero-inferior view. Note the profuse supply to the ventricles, especially the left, and the sparse supply to the atria (i.e. above the main trunks of the right and left coronary arteries (RCA, LCA) situated in the atrioventricular grooves).

Coronary insufficiency

If a coronary artery becomes narrowed or blocked, the heart muscle becomes anoxic, particularly on exertion, and this causes chest pain **'angina'**. The most common cause is atherosclerosis—deposits of cholesterol in the walls on which thrombi (blood clots) can develop, and which become calcified. If death of cardiac muscle tissue (**myocardial infarction**) occurs, the patient is said to have had a 'coronary' or 'heart attack'.

It is often possible to bypass the obstruction and restore coronary circulation by transplanting veins from another part of the body. Alternatively, a flexible cannula with an inflatable end-piece can be inserted into the femoral artery, guided radiologically up the aorta and into the coronary artery, and there inflated to crack the atherosclerotic plaque and restore patency to the artery.

Drugs that prevent clotting or dissolve clots are also used. If, after a myocardial infarction, an affected papillary muscle should snap, then sudden incompetence of the affected a-v valve will develop and the efficiency of the pumping action of the heart is suddenly reduced.

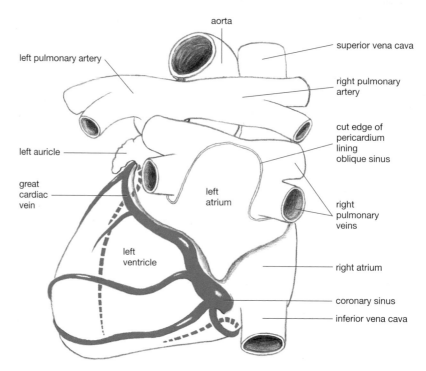

5.4.14 The venous drainage of the heart.

Innervation and conducting system of the heart

The heart beats continuously and autonomously throughout life, the rate of contraction and stroke volume being adjusted to physiological demand.

The rate of contraction is determined by the **sinu-atrial** (S-A) **node** (5.4.15), a group of specialized muscle cells that act as a 'pacemaker'. The S-A node is situated at the upper end of the crista terminalis, but cannot be detected by eye alone. Its intrinsic activity is modified by **vagal** (parasympathetic preganglionic) and **sympathetic** (postganglionic) fibres, which arise from the cardiac plexuses beneath the arch of the aorta; the vagal fibres end on ganglion cells near the node.

From the S-A node, impulses pass via the atrial muscle to reach the **atrioventricular (A-V) node**. This node (which also is only detectable histologically) lies to the left of the inferior vena caval opening on the atrial side of the atrioventricular septum; it receives largely sympathetic fibres.

The muscle of the atria is separated from that of the ventricles by the fibrous tissue skeleton, which also gives attachment to the atrioventricular valves. The fibrous tissue prevents electrical activation of the atria from spreading to the ventricles except by the **atrioventricular bundle** (of His), a band of specialized cardiac muscle fibres which arises from the A-V node and passes through the fibrous skeleton, transmitting electrical activation from the A-V node to the ventricular muscle.

The A-V bundle passes deep to the septal cusp of the tricuspid valve to reach the membranous part of the interventricular septum where it divides into **right** and **left bundles**. The **right bundle** passes across a muscular 'bridge' (the **moderator band**) to the anterior wall of the right ventricle where, like the corresponding **left bundle** which passes down the left side of the interventricular septum, it breaks into **Purkinje fibres**, which terminate among the ventricular muscle fibres.

Qu. 4K *A 55-year-old man complains of the sudden onset of severe central chest pain which radiated to the inner aspect of his left arm. A coronary angiogram revealed a marked narrowing of the anterior interventricular branch of the left coronary artery. In addition it was found that the atria and ventricles were not beating synchronously. How can you explain these symptoms and signs?*

The postganglionic sympathetic fibres to the heart originate in the cervical and upper thoracic sympathetic ganglia and the cardiac plexus (p. 18); their preganglionic sympathetic neurons are situated in the upper part of the thoracic spinal cord (T1–T5). The postganglionic parasympathetic neurons are situated in the wall of the heart and receive preganglionic fibres from the vagus nerves.

The pericardium

The heart and its blood vessels are enclosed in a thick fibrous sac, the **fibrous pericardium**. This is attached inferiorly to the margin of the central tendon of the diaphragm and superiorly to the outer coats of the great vessels. It protects and stabilizes the heart within the thoracic cavity and limits the excursions of the diaphragm. It prevents acute overdistension of the heart (e.g. during exercise) but, if the volume of the heart is consistently increased, will grow to accommodate it (see **5.4.6**).

The pericardium is innervated by sensory fibres of the **phrenic nerves** (C3, 4, 5) which pass downward on either side of the sac (**5.4.8**) to reach the diaphragm. Small branches of the internal thoracic arteries run with the phrenic nerves to supply the pericardium.

Within the fibrous pericardium is a double-layered sac of **serous pericardium**. Its parietal layer lines the fibrous pericardium and its visceral layer almost entirely covers the surface of the heart. The two layers are continuous around the great arteries and great veins which leave and enter the heart (**5.4.16**). A potential **pericardial cavity** containing a thin film of fluid is present between the serous layers to allow the heart to beat freely within the fibrous pericardium.

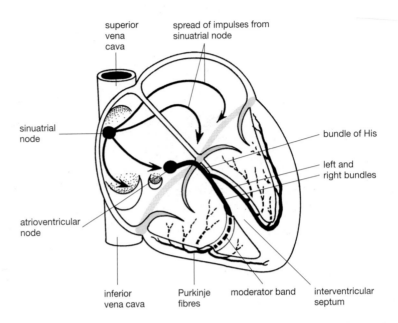

5.4.15 The conducting system of the heart. The fibrous skeleton of the heart separates the muscle of the atria from that of the ventricles, which are connected only by the bundle of His.

Abnormal heart beats

If the conducting system is disrupted, either congenitally or by a myocardial infarction, then the atria beat at one speed and the ventricles at their own, slower speed ('heart block'). Furthermore, the ventricular contraction rate does not increase on exertion.

Sometimes the atrial muscle does not contract synchronously (atrial fibrillation). This leads to irregular contraction of the ventricles, and may be corrected by administering a synchronizing electrical shock to the heart.

Pain from the heart

The sensory fibres that convey the pain that results from ischaemia of the heart muscle (**angina**) pass to the upper thoracic spinal cord with the sympathetic nerves. Pain from the heart is therefore referred to the skin innervated by T1–T5 and is sensed as a very severe, constricting central chest pain, which can spread to the inner aspect of the left arm.

Between the great arteries (aorta, pulmonary artery) and the superior vena cava is the transverse sinus of the pericardium, a space created by the breakdown of the embryonic dorsal mesocardium. The pouch of serous pericardium on the posterior wall of the heart between the orifices of the great veins is called the oblique sinus.

Imaging the heart

CT and MRI

The chambers of the heart and major vessels can all be identified on CT and MRI images of the chest. Identify them on **5.4.17**, **5.3.13a**, and the through-thorax series **5.6.1**.

Ultrasound imaging

Ultrasound is increasingly used to image the heart, especially in the fetus. It is non-damaging and enables the movement of the heart valves to be viewed in 'real-time'. However, the images are not easy to interpret.

Pericarditis

Inflammation of the pericardium (pericarditis) can cause an effusion of fluid to collect in the pericardial cavity. If a significant amount of fluid accumulates, the fibrous pericardium becomes distended and the action of the heart becomes less efficient because its venous filling is impeded (cardiac tamponade). Such a large effusion can be drained by inserting a needle upward and backward between the xiphoid process and the costal margin on the left side, through the 'bare area' of the pericardium (that part not covered anteriorly by pleura).

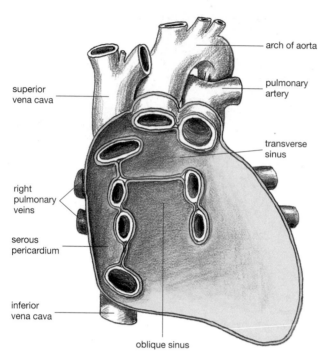

5.4.16 Posterior wall of serous pericardial sac to show the entrance of vessels and the transverse (T) and oblique (O) sinuses.

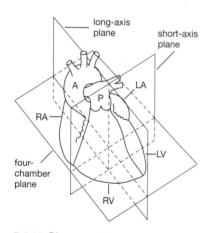

5.4.18 Diagram to show the principal axes of the heart used in ultrasound examination. Right atrium (RA), right ventricle (RV), pulmonary trunk (P), left atrium (LA), left ventricle (LV), aorta (A).

5.4.18 illustrates the three major planes used to examine the heart by ultrasound. 5.4.19 shows a 'parasternal long axis' view in which the papillary muscles, chordae tendineae, and anterior and posterior leaflets of the mitral valve can be seen. Note the proximity of the attachment of the anterior leaflet of the mitral valve to that of the non-coronary (left posterior) cusp of the aortic valve, which is open. Ultrasound can also visualize the leaflets of the aortic valve in systole and diastole and, combined with Doppler imaging, show flow through the chambers, vessels, and any defects.

Questions and answers

Qu. 4A *In which direction would blood tend to flow through a defect in the interventricular septum?*

Answer If the defect is between the ventricles, the greater pressure in the left ventricle would force arterial blood into the right ventricle and overload it.

Qu. 4B *If the pulmonary artery is very narrow, how could blood reach the lungs after the infant is born?*

Answer Blood could continue to reach the lungs provided that the ductus arteriosus did not shut. As the blood would be less oxygenated than normal, the ductus is unlikely to close.

Qu. 4C *What other relatively common congenital abnormalities of the heart would you expect from a knowledge of its development?*

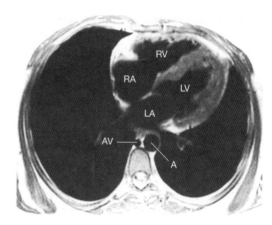

5.4.17 Axial MRI showing the four chambers of the heart. Right atrium (RA); right ventricle (RV), left atrium (LA); left ventricle (LV); aorta (A), azygos vein (AV).

Answer Other common cardiac anomalies include:

- **interatrial septal defect**, usually failure of the foramen ovale to close, but occasionally failure of foramen primum closure;
- **interventricular septal defect**, in either its muscular or membranous (endocardial cushion-derived) part;
- **failure of septation of the great vessels**, including transposition of the aorta and pulmonary artery;
- **failure of the ductus arteriosus to close after birth**.

Qu. 4D *Examine radiograph 5.4.6. What abnormality is visible?*

Answer The heart is enlarged, its transverse diameter being $0.6 \times$ that of the thorax.

Qu. 4E *Examine radiograph 5.4.7. What developmental anomaly is present?*

Answer The heart apex points to the right. Also the bubble of gas in the stomach can just be seen beneath the right dome of the diaphragm, and the left dome is higher than the right (presumably over the liver). It is therefore likely that the entire body is left–right reversed ('situs inversus').

Qu. 4F *What is the function of the chordae tendineae and papillary muscles?*

Answer By anchoring the valvular cusps, the chordae tendineae, attached to papillary muscles, ensure that the atrioventricular valves are not forced back into the atria during ventricular systole. They therefore prevent atrioventricular valve incompetence.

Qu. 4G *If the tricuspid valve cannot close properly during systole, what will be the effect on the right atrium? If the defect is considerable, what will be the effect on the liver and the lower extremities?*

Answer The right atrium will receive blood not only from the venae cavae but also from the right ventricle.

This will lead to inefficient emptying of the venous system, causing enlargement of the liver and swelling of the legs due to a build-up of subcutaneous tissue fluid (oedema); the fluid is insufficiently drained because of the higher hydrostatic pressure in the veins and thus the capillaries.

Qu. 4H *Why is the muscular wall of the left ventricle normally three times thicker than that of the right ventricle?*

Answer Greater pressure must be exerted by the left ventricle in order to pump oxygenated blood into the aorta and around the body, than by the right ventricle which pumps blood through the pulmonary circulation. The thickness of the ventricular walls reflects this difference.

Qu. 4I *If the mitral valve became narrowed (stenosed) what would be the effect on the heart and lungs?*

Answer The pumping of the left side of the heart would become inefficient, the left atrium dilated; the lungs would become congested with blood and the alveoli oedematous, leading to inefficient oxygenation of the blood. The congestion would also place a greater load on the right ventricle which, in turn, might fail.

Qu. 4J *If a person was found to have a dilated left atrium, what abnormality of a heart valve would you expect? What radiological view of the heart would most easily reveal this condition?*

Answer An enlarged left atrium suggests either mitral stenosis or incompetence. The left atrium lies anterior to the oesophagus and when enlarged can indent the oesophagus. This can be seen on a lateral radiograph of a barium swallow or a CT or MRI scan.

Qu. 4K *A 55-year-old man complains of the sudden onset of severe central chest pain which radiated to the*

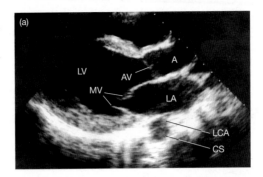 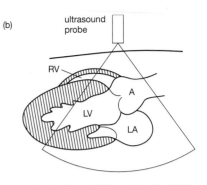

5.4.19 (a) Heart ultrasound, parasternal long-axis view during ventricular diastole. The mitral valve (MV) is open, to allow blood to enter the left ventricle (LV) from the left atrium (LA). The aortic valve (AV) is closed, preventing reflux from the aorta (A). Cross-sections of the left coronary artery (LCA) and coronary sinus (CS) are seen in the atrio-ventricular groove. (b) Explanatory diagram; RV, right ventricle.

inner aspect of his left arm. A coronary angiogram revealed a marked narrowing of the anterior interventricular branch of the left coronary artery. In addition it was found that the atria and ventricles were not beating in synchrony. How can you explain these symptoms and signs?

Answer The anterior interventricular branch of the left coronary artery was shown to be blocked. Cardiac muscle in the anterior wall of both right and left ventricles would be deprived of blood. The artery also supplies the interventricular septum, and the conducting system of the ventricles (atrio-ventricular bundle (of His)) may be affected. Anoxic cardiac muscle causes central chest pain which classically radiates down the medial aspect of the left arm. Depending on the degree of damage to the atrioventricular bundle, the ventricles may beat more slowly than the atria because they have been disconnected from the atrial pacemaker and beat with their own rhythm. This leads to inefficient pumping.

Mediastinum and paravertebral area

Mediastinum and paravertebral area

The mediastinum is situated centrally in the thorax, between the right and left lungs in their pleural cavities, and extends between the neck and the abdomen. Its large central compartment (middle mediastinum) houses the heart and pericardium. The space in front of the heart (anterior mediastinum) is very narrow. The compartments above the heart (superior mediastinum) and behind the heart (posterior mediastinum) contain all the other structures that enter and leave the chest. Structures entering or leaving the chest through the superior mediastinum must pass through the narrow space between the apices of the pleural cavities with their covering suprapleural membranes; inferiorly they must pass through or behind the diaphragm. Laterally, the superior and posterior mediastinum extend into the paravertebral area behind the medial parts of the pleural cavities.

The anterior mediastinum

The **anterior mediastinum** (5.5.1) is the narrow space between the pericardium and the anterior chest wall. It contains a little fatty connective tissue within which are:

- The right and left **internal thoracic vessels**. The arteries are branches of the first part of the subclavian artery. They arch forward over the anterior part of the suprapleural membrane and then pass downward behind the costal cartilages, giving off branches to each intercostal space, before dividing into the superior epigastric and musculophrenic arteries (p. 51). The veins drain upward to join the subclavian veins.
- Internal thoracic **lymphatics**. This chain of intercommunicating vessels and nodes drains the anterior chest wall, including the medial parts of the **breast**, and communicates with lymphatics of the anterior abdominal wall. Tumours from the breast can therefore spread into the chest cavity via this route.
- The **thymus**, which can extend into the anterior mediastinum from the superior (see below), especially in children.

  The anterior mediastinum may also contain parts of the parathyroid or thyroid glands if they have descended abnormally low during development, associated with the descent of the thymus.

The superior mediastinum

The superior mediastinum (5.5.2) is the space behind the manubrium and above the heart, i.e. between the narrow inlet to the thorax and the (larger) plane which extends horizontally from the sternal angle (manubriosternal joint) to the lower part of the body of T4 vertebra (plane of Louis). In it are found:

- the trachea and oesophagus;
- the arch of the aorta and its branches—the brachiocephalic trunk, the left common carotid artery and the left subclavian artery;
- the superior vena cava; its major tributaries, the left and right brachiocephalic veins;
- the thoracic duct and bronchomediastinal lymph trunks and their associated lymph nodes;
- the upper parts of the thymus;
- parts of the phrenic and vagus nerves and the sympathetic chain.

  The **trachea** is continuous with the larynx at the level of the sixth cervical vertebra. It lies anterior to the oesophagus and passes down in the midline to enter the superior mediastinum behind the manubrium, then, inclining backward and slightly to the right, ends behind the sternal angle by dividing into the right and left main bronchi (p. 57, 60), which pass to the roots of the lungs. At its lower end it lies behind the arch of the aorta, the brachiocephalic and left common carotid arteries ascending in a V around it. Its upper part is covered anteriorly by the pretracheal fascia and crossed by the left brachiocephalic vein.

  The trachea is supplied by a fine longitudinal anastomosis formed by small branches of nearby arteries. Extensive surgical dissection around the trachea can destroy this network with serious consequences.

  The **arch of the aorta** is the continuation of the ascending aorta. It passes backward and to the left, arching over the right pulmonary artery and left main bronchus, then bends downward into the posterior mediastinum, where it continues as the descending

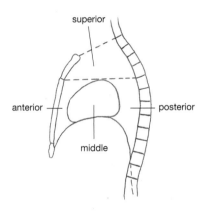

5.5.1 The mediastinum and its divisions.

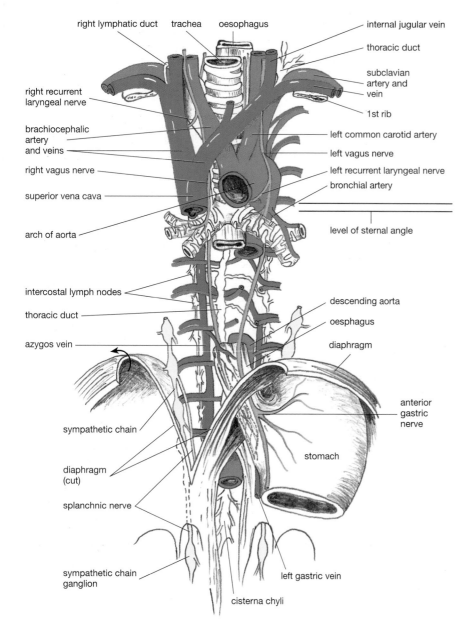

5.5.2 Posterior and superior parts of the mediastinum seen after removal of the heart and splitting of the diaphragm.

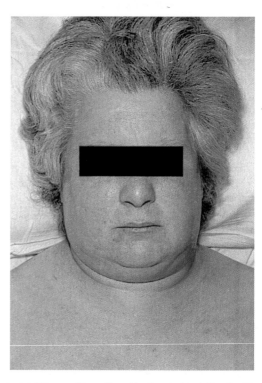

5.5.3 Woman with swelling of the face and neck. See Qu. 5A.

- The **left common carotid artery** passes upward into the neck on the left side of the trachea.
- The **left subclavian artery** has a longer course than that on the right, but leaves the thorax to enter the axilla in the same way.

A small artery to the thyroid (thyroidea ima) may ascend from the arch of the aorta anterior to the trachea (with the inferior thyroid veins; see below).

The **left** and **right brachiocephalic veins** are each formed behind the sternoclavicular joint and in front of their respective arteries by the union of subclavian vein and the internal jugular vein. The left brachiocephalic vein passes obliquely downward and to the right behind the manubrium to join the right brachiocephalic veins. Together they form the **superior vena cava**, which passes downward, is joined by the azygos vein (which has arched forward over the root of the right lung) and then enters the pericardium to end in the right atrium.

Veins draining the isthmus of the thyroid (inferior thyroid veins) descend in the pretracheal fascia to enter the left brachiocephalic vein; they can be a hazard during an emergency operation to open the trachea. A further hazard of this operation in young children is that the left brachiocephalic vein is situated above, rather than behind the manubrium.

Qu. 5A *A woman (5.5.3) develops increasing swelling that is restricted to her face and neck. What is a likely cause?*

The **phrenic nerves** supply the diaphragm and its coverings of pleura and peritoneum; they also supply

thoracic aorta. In elderly people the walls of the arch are less elastic and so it bulges more to the left.

The inferior aspect of the arch is attached to the bifurcation of the pulmonary trunk by a fibrous cord, the **ligamentum arteriosum**, the remnant of the embryonic ductus arteriosus (p. 69). From the arch arise three main branches:

- The **brachiocephalic** (innominate) **artery** which passes up and to the right of the trachea to a position behind the sternoclavicular joint, where it divides into the right subclavian artery and the right common carotid artery. The right subclavian artery leaves the thorax by arching over the suprapleural membrane and first rib to enter the axilla; the right common carotid artery passes upward into the neck on the side of the trachea.

sensory fibres to the pericardium and mediastinal pleura. They originate on either side of the neck from spinal nerves C3–C5. They run downward over the scalenus anterior muscle and enter the superior mediastinum between the subclavian artery and vein.

The **right phrenic nerve** runs down over the venous structures that form the right side of the superior and middle mediastinum (the right brachiocephalic vein, superior vena cava, right atrium and inferior vena cava within the pericardium) to pierce the diaphragm.

The **left phrenic nerve** passes down through the chest, lying on arterial structures that form the left side of the superior and middle mediastinum (the left subclavian artery, arch of the aorta, left pulmonary artery, and the left ventricle within the pericardium) before it pierces the diaphragm.

Qu. 5B *What will be the effect if the phrenic nerve is (a) cut in its course through the middle mediastinum; (b) irritated by inflammation of the diaphragm?*

The **vagus nerves** (X cranial nerves) of both sides provide the preganglionic parasympathetic and sensory supplies to the respiratory tract, heart, and all the derivatives of the fore- and mid-gut within the abdomen. They originate from the brainstem within the skull and descend through the neck, lying behind the internal and common carotid arteries and internal jugular vein in a sheath of connective tissue (the carotid sheath; Vol. 3, Chapter 6, Section 10) to enter the superior mediastinum.

The **right vagus** passes downward over the subclavian artery to enter the superior mediastinum where, lying lateral to the trachea, it passes behind the root of the right lung.

As it crosses the subclavian artery it gives off the **right recurrent laryngeal nerve**, which loops beneath the artery and then runs upward between the trachea and oesophagus to supply the larynx (Vol. 3, Chapter 6, Section 6).

The **left vagus nerve** enters the superior mediastinum behind the left common carotid artery and, passing downward, crosses to the left side of the arch of the aorta to gain the posterior aspect of the root of the left lung. As it crosses the aorta, it gives off the **left recurrent laryngeal nerve**, which loops beneath the ligamentum arteriosum then ascends to the larynx in the groove between the trachea and oesophagus. If the arch of the aorta becomes distended by an aneurysm, this can stretch the recurrent laryngeal nerve, causing a hoarse voice.

At the roots of the lungs, the vagus nerves supply branches to the **pulmonary plexus** and, passing medially on to the oesophagus, form the **oesophageal plexus**. Before passing through the diaphragm, however, fibres of the left and right vagus nerves reaggregate to form, respectively, the **anterior** and **posterior gastric nerves**.

The vagus nerves also supply cardiac (depressor) nerves to the heart, but these arise mostly in the neck, reflecting the origin of the heart ventral to the pharyngeal arches. They pass down through the superior mediastinum, form a **cardiac plexus** beneath the arch of the aorta, and are distributed almost entirely to the sinu-atrial node.

Qu. 5C *What would be the effects on the heart and lungs of cutting the left vagus nerve in the neck?*

The **thymus gland** is one of the primary lymphoid organs. It increases in size to birth, remains large for about 30 years, then atrophies and becomes infiltrated with (and difficult to distinguish from) fat. It consists of a right and a left lobe, which are situated behind the sternum. When large, it covers the great vessels and upper part of the pericardium and can extend into the neck.

The thymus is crucial to the cellular immune response. Its cells 'process' lymphocyte stem cells (which have been produced in the bone marrow) to produce and maintain clones of immunologically competent T (thymic) lymphocytes. The thymus also produces a number of hormones and cytokines which have immunomodulatory effects.

The other structures in the superior mediastinum (oesophagus, thoracic duct and other lymph ducts and nodes, elements of the sympathetic nervous system) are also present to a greater or lesser extent in the posterior mediastinum (see below).

The posterior mediastinum

The **posterior mediastinum** (5.5.2) is the space below the superior mediastinum (i.e. below the plane of Louis) and behind the pericardium, extending downward behind the sloping posterior part of the diaphragm. It extends laterally behind the pleural cavities into the paravertebral area. The region contains:
- the oesophagus;
- the descending aorta and its branches;
- the azygos vein and its tributaries;
- the thoracic lymph duct and associated nodes;
- the thoracic sympathetic chain and its branches.

The **oesophagus** is the downward continuation of the pharynx. Like the trachea, it begins in the midline of the neck at the level of the sixth cervical vertebra. Follow it down through the series of CT scans of the thorax (5.6.1). It enters the thoracic inlet in the narrow space between the body of the first thoracic vertebra and the trachea, slightly to the left of the midline, and descends behind the trachea and pericardium, to the right of the aortic arch, lying on the middle thoracic vertebrae.

Qu. 5D *Which chamber of the heart lies immediately anterior to the oesophagus?*

Qu. 5E *Which pulmonary artery crosses anterior to the oesophagus?*

As it approaches the diaphragm, the oesophagus inclines to the left, crossing anterior to the aorta. It then enters the abdomen by piercing the crura a little (2 cm) to the left of the midline at the level of the tenth thoracic vertebra and, after a short intra-abdominal

Thymus, and ectopic glands in the anterior mediastinum

The thymus, thyroid, and parathyroid glands all develop from the primitive pharyngeal endoderm (Vol. 3, p. 38). The thymus normally migrates downward the furthest and may carry parathyroid glands (particularly the inferior) with it. Tumours of any of these organs can involve the superior mediastinum and can compress the trachea and other structures in the narrow inlet to the thorax. Thymic tumours are particularly associated with autoimmune diseases, such as myasthenia gravis (auto-antibodies to the neuromuscular acetylcholine receptor).

Because the thymus is essential to the normal development of lymphoid tissues, removal of the thymus in early life leads to progressive wasting of peripheral lymphoid organs and inability to mount an effective immune response; this usually proves fatal.

Infection of the mediastinum

The posterior mediastinum can become infected (mediastinitis) if a sharp swallowed object such as a fish-bone pierces the wall of the oesophagus; infection can also track down into the posterior mediastinum from the retropharyngeal space if there is a perforation of the pharynx in the neck.

Congenital oesophageal abnormalities

The oesophagus develops from the endoderm of the foregut and its surrounding mesenchyme. During development it elongates considerably as the stomach moves to its definitive position. It also rotates to the right with the stomach, causing the right vagus nerve to lie posterior to the lower end of the oesophagus.

The separation of the trachea from the oesophagus has already been described (p. 55). If the tracheo-oesophageal septum, which normally separates the trachea from the oesophagus, does not develop properly, a **tracheo-oesophageal fistula** is formed. Milk swallowed by the newborn child can then enter the lungs, with serious consequences. Tracheo-oesophageal fistula is often associated with **oesophageal atresia**, in which the oesophageal lumen becomes obliterated during development. Oesophageal diverticula can also occur. The cause of these abnormalities is unclear.

During fetal life oesophageal atresia causes an accumulation of amniotic fluid (polyhydramnios), because swallowed amniotic fluid is normally absorbed by the gut and excreted via the placenta. It is obviously important that both tracheo-oesophageal fistula and oesophageal atresia are corrected soon after birth.

Congenital **hiatus hernia**, in which part of the stomach obtrudes through the oesophageal hiatus into the thoracic cavity, can be associated with failure of the oesophagus to lengthen properly, or a developmental defect in the diaphragm.

Diaphragmatic (hiatus) hernia

Not uncommonly, the lower end of the oesophagus and part of the fundus of the stomach slide upward through the oesophageal opening of the diaphragm into the thorax (**hiatus hernia**). Many cases are recognized only as a result of routine imaging, but 'heartburn'—lower central chest pain which, as its name implies, can be difficult to distinguish from heart pain—can result from reflux of gastric acid on to the oesophageal mucosa. The pain can radiate to the neck or the back. Larger hiatus hernias can cause breathlessness by decreasing the volume of the thorax.

course, joins the stomach at the cardiac sphincter (p. 136).

The oesophagus is supplied by small branches arising directly from the descending aorta. Its lower end also receives blood from the left gastric artery, which arises in the abdomen and sends an ascending branch to the lower part of the oesophagus via the oesophageal opening in the diaphragm. Similarly, oesophageal venous blood passes largely to the azygos system of veins. At the lower end, however, it passes downward into the left gastric vein which joins the hepatic portal circulation.

The lower end of the oesophagus is therefore an important site of portal–systemic anastomosis. Thin-walled veins joining the two systems form vertical channels just beneath the mucosa of the oesophagus. If excessive pressure develops in the hepatic portal vein (portal hypertension; p. 163), the varicose veins can rupture, causing catastrophic bleeding.

Peristalsis in the oesophagus is controlled largely by an intrinsic plexus of nerves and initiated by the swallowing reflex. The oesophagus does, however, also receive postganglionic sympathetic fibres, and the oesophageal plexus, derived from the vagus nerves, lies on its lower part. Sensory fibres pass with the sympathetic innervation to the upper thoracic cord. Stimulation of pain fibres, for instance by reflux of gastric acid into the lower end of the oesophagus, causes heartburn (see below).

At the upper end of the oesophagus the muscle creates a zone of high pressure (the upper oesophageal sphincter), which prevents air from entering the oesophagus and oesophageal contents from refluxing into the pharynx. When food is swallowed, the 'upper oesophageal sphincter' relaxes and sequential peristaltic waves convey the food or fluid down the oesophagus, through the lower oesophageal sphincter (p. 136) and into the stomach. Although swallowing is said to consist of oral, pharyngeal, and oesophageal phases (Vol. 3, Chapter 7), it is a smooth, co-ordinated movement, voluntarily initiated but continued automatically. Difficulty in swallowing (dysphagia) is often caused by either neural or mechanical (e.g. tumours) disorders. It can be investigated by the patient drinking a small amount of a barium-containing fluid, the progress of which is followed by radiography (5.5.4), or, alternatively, by endoscopy (see 6.4.6).

The **descending thoracic aorta** (5.5.2, 5.5.5) is the continuation of the arch of the aorta at the left side of the fourth thoracic vertebra. It passes downward and gradually to the right, to lie on the anterior aspect of the lower thoracic vertebrae, and behind the oesophagus. It then leaves the thorax by passing behind and between the muscular crura of the diaphragm at the level of the twelfth thoracic vertebra.

It supplies **posterior intercostal arteries** to the lower nine intercostal spaces; those on the right cross the anterior aspect of the thoracic vertebrae to reach the right side of the chest. Other small branches supply arterial blood to the bronchi (**bronchial arteries**) and oesophagus.

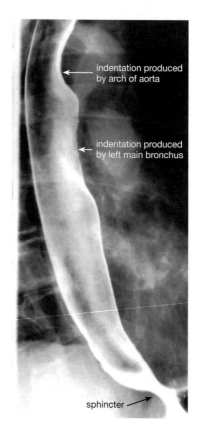

indentation produced by arch of aorta

indentation produced by left main bronchus

sphincter

5.5.4 Double-contrast ('barium swallow') radiograph of oesophagus in oblique view.

The **azygos vein** (5.5.2, 5.5.5) drains most of the posterior chest wall. It lies on the right of the lower thoracic vertebral bodies (azygos = unpaired) and receives directly the veins draining the intercostal spaces on the right. On the left side, these veins form upper and lower hemiazygos veins which cross the midline behind the aorta to reach the azygos vein. The azygos vein ends by arching forward over the root of the right lung to drain into the superior vena cava just before it enters the pericardial sac.

The **thoracic duct** is a large lymphatic channel, arising in the abdomen from a lymphatic sac, the **cisterna chyli**, which lies on the anterior aspect of the first lumbar vertebra. The cisterna chyli receives lymph from almost every part of the body below the diaphragm, including the abdominal and pelvic viscera (the chylomicron-rich lymph from the small bowel gives the sac its name), the abdominal walls, and the lower limbs.

The thoracic duct ascends from the abdomen lying behind the aorta and crura of the diaphragm. As it passes up in front of the thoracic vertebrae it lies behind the oesophagus. At about the mid-thorax it crosses gradually to the left of the midline as it ascends. It then lies to the left of the oesophagus and eventually passes laterally and forward at the root of the neck, arching over the left subclavian artery, to drain into the junction of the left internal jugular and left subclavian veins.

splanchnic nerve) to supply directly the catecholamine-secreting cells of the adrenal medulla (p. 223).

At the upper end of the thoracic cavity, the sympathetic chain lies on the neck of the first rib. Here, the first thoracic ganglion is often fused with the inferior cervical

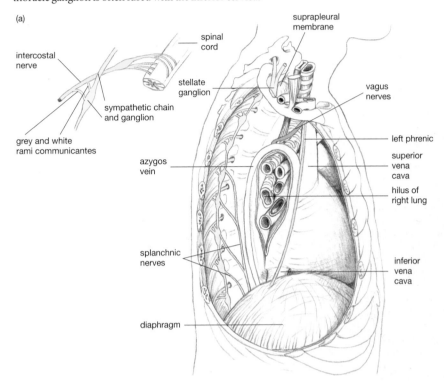

(a)

intercostal nerve

spinal cord

stellate ganglion

suprapleural membrane

sympathetic chain and ganglion

vagus nerves

grey and white rami communicantes

azygos vein

left phrenic

superior vena cava

hilus of right lung

splanchnic nerves

inferior vena cava

diaphragm

Damage to the thoracic duct

The thoracic duct has many minute anastomoses with other lymph vessels, and so can be ligated without embarrassing the return of lymph to the bloodstream. If, however, the duct wall becomes damaged by disease, the 2 litres per day of protein-rich, milky lymph that it carries can leak out into the pleural cavity (chylothorax).

Carcinoma of the lung apex

A tumour of the apex of the lung may spread to involve the sympathetic chain at the neck of the first rib. This can disrupt the sympathetic supply to that side of the head and neck (see p. 18). The resultant group of signs (**Horner's syndrome**), which you should be able to explain when you have studied the head and neck (Vol. 3), are flushed dry skin, drooping eyelid, constricted pupil, and a sunken eyeball on the affected side.

The thoracic duct receives lymph from the left side of the thoracic cage and much of the thoracic contents. Lymph from the right side of the thoracic cage, and some lymph from the right lung and right side of the mediastinum, drains into a small **right lymphatic duct** which runs up the right side of the mediastinum, to end, like the thoracic duct, in the junction of the (right) subclavian and internal jugular veins.

Qu. 5F *Why is the thoracic duct lymph milky in appearance, particularly after meals?*

The paravertebral area

The principal feature of the paravertebral area is the **thoracic sympathetic chain** and its branches (5.5.5). This lies vertically on the necks of the ribs on either side of the thoracic vertebral column. It comprises a series of about 12 **sympathetic ganglia** linked together by pre- and postganglionic fibres. The chain of ganglia lies directly anterior to the anterior primary rami of the thoracic spinal nerves as they pass into the intercostal spaces.

The ganglion cells receive preganglionic sympathetic axons (white rami communicantes) of cell bodies situated in the lateral horn of the T1–L2 segments of thoracic grey matter, and send postganglionic axons (grey rami communicantes) into the intercostal nerves for distribution to the chest and abdominal walls.

Some preganglionic fibres do not synapse in the ganglia but pass through them to form medial branches (**thoracic splanchnic nerves**; the *greater* from ganglia T5–T9, the *lesser* from T10 and T11; and the *least* from ganglion T12). These descend on the sides of the thoracic vertebrae and pierce the crura of the diaphragm to supply the gut via synapses in the abdominal sympathetic ganglia (p. 18), and (via the least

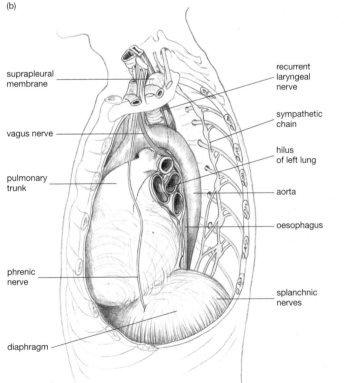

(b)

suprapleural membrane

recurrent laryngeal nerve

vagus nerve

sympathetic chain

pulmonary trunk

hilus of left lung

aorta

oesophagus

phrenic nerve

splanchnic nerves

diaphragm

5.5.5 Interior of (a) right and (b) left sides of the chest cavity to show the sympathetic chain and splanchnic nerves.

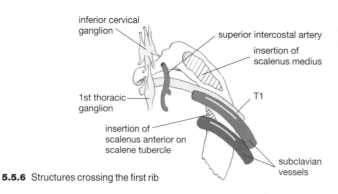

5.5.6 Structures crossing the first rib

inferior cervical ganglion
superior intercostal artery
insertion of scalenus medius
1st thoracic ganglion
T1
insertion of scalenus anterior on scalene tubercle
subclavian vessels

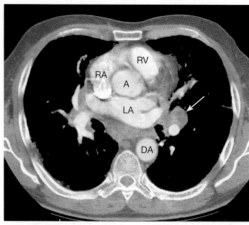

5.5.9 CT through posterior mediastinum. See Qu. 5H.

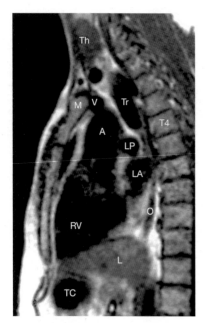

5.5.7 Sagittal near midline MRI, showing the structures in the mediastinum. Arch of aorta (A), liver (L), left atrium (LA), left pulmonary artery (LP), manubrium (M), oesophagus (O), fourth thoracic vertebra (T4), transverse colon (TC), thyroid (Th), trachea (Tr), left brachiocephalic vein (V).

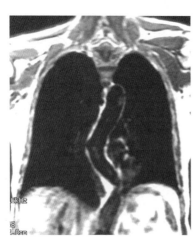

5.5.8 Coronal MRI showing a distorted descending aorta.

ganglion to form a 'stellate' ganglion. Lateral to the ganglion (5.5.6) lie:

• the superior intercostal artery (a branch of the costo-cervical trunk of the subclavian artery, which supplies the upper two intercostal spaces);
• the anterior primary ramus of T1, which leaves the thorax to become the lowest root of the brachial plexus (Vol. 1, p. 111);
• the subclavian artery and scalenus anterior.

Qu. 5G *If a carcinoma involving the apex of the lung also involved the T1 root of the brachial plexus, what would be the effects in the upper limb?*

Imaging

Radiology

On plain P-A chest radiographs the trachea can be identified readily as it enters the thorax. The arch of the aorta forms a 'knuckle' to the left of the midline, which becomes increasingly prominent with age as the aorta loses elasticity.

The oesophagus is best viewed on a lateral radiograph after swallowing a small amount of radiopaque material ('barium swallow'; 5.5.4).

The major vessels of the mediastinum can be outlined by angiography (arteriography, venography, lymphangiography).

Computed imaging

CT and MR images show clearly the trachea, oesophagus, and major vessels. Identify them on the CT series through the chest shown in 5.6.1 and MRIs shown in 5.5.7 and 5.6.2. 5.5.8 is a coronal MRI showing a descending aorta that is rather distorted, perhaps by atherosclerosis.

Qu. 5H *What abnormality do you see in 5.5.9, and what might be the origin of the abnormality?*

Questions and answers

Qu. 5A *A patient (5.5.3) develops increasing swelling that is restricted to her face and neck. What is a likely cause?*

Answer Pressure on the superior vena cava will obstruct venous drainage from the head and neck to the right atrium and will therefore cause swelling of the face and neck.

Qu. 5B *What will be the effect if the phrenic nerve is (a) cut in its course through the middle mediastinum; (b) irritated by inflammation of the diaphragm?*

Qu. 5H *What abnormality do you see in 5.5.9, and what might be the origin of the abnormality?*

Answer (a) The dome of the diaphragm will be paralysed on the affected side. (b) Referred pain will be felt in the tip of the shoulder due to stimulation of C4 which supplies the diaphragm (phrenic nerve) and also the skin of the shoulder (supraclavicular nerves).

Qu. 5C *What would be the effects on the heart and lungs of cutting the left vagus nerve in the neck?*

Answer Cutting the left vagus nerve would increase the heart rate and reduce bronchial secretion. It would also decrease gastrointestinal tract movements and secretion in the abdomen.

Qu. 5D *Which chamber of the heart lies immediately anterior to the oesophagus?*

Answer The left atrium.

Qu. 5E *Which pulmonary artery crosses anterior to the oesophagus?*

Answer The right pulmonary artery crosses the oesophagus and indents it.

Qu. 5F *Why is the thoracic duct lymph milky in appearance, particularly after meals?*

Answer Because the lacteals drain small particles of absorbed triglyceride (chylomicrons) from the intestinal villi.

Qu. 5G *If a carcinoma involving the apex of the lung also involved the T1 root of the brachial plexus, what would be the effects in the upper limb?*

Answer A carcinoma involving the T1 spinal nerve (which forms the lowest root of the brachial plexus) would cause wasting of the small muscles of the hand on the affected side. This would interfere with fine manipulative movements of the fingers (e.g. in handwriting).

Qu. 5H *What abnormality do you see in 5.5.9, and what might be the origin of the abnormality?*

Answer 5.5.9 shows large abnormal masses in the hilar regions of the lungs. They are enlarged lymph nodes. The likely origin is secondary deposits from a bronchial carcinoma.

Sectional anatomy of the thorax revealed by CT and MRI

Sectional anatomy of the thorax revealed by CT and MRI

Increasingly, diagnosis of chest disease involves imaging the chest by CT or MRI. CT can be adjusted to show either the bones and most soft tissues well, or to show the structure of the lungs and airway. MRI is used primarily for the heart and other soft tissues. The images are sections which can be computed in any orientation, although axial (horizontal) sections and vertical sections are the most commonly used. The student therefore needs to be able to interpret sectional views as well as developing an understanding of the anatomy of the chest that is built up from studies of the living chest, of dissected cadavers, of conventional radiographs and of other images. Remember that all axial sections are presented as if the view is from the patient's feet, so that the patient's right side is the left side of the illustration.

The set of spiral computed tomograms (5.6.1b–l) are axial sections through the chest at progressively more caudal intervals, illustrated in (5.6.1a). These images are optimized for the soft tissues. The subject had inhaled and held a very deep breath during the exposure, so that the rib cage is elevated (the manubriosternal joint is opposite T3 rather than T4); and the air-filled (black) lungs extend down into the depths of the pleural sacs. In addition, in this subject, the hepatic flexure of the colon is unusually high.

As an aid to learning, place a cover across the adjacent labelled diagram and discover how many structures you can identify, then check that your observations are correct. Compare these images with 5.3.13b optimized for the lung tissue.

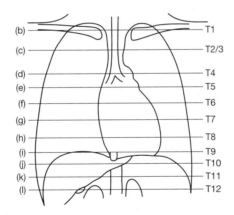

(a)

(b) — T1
(c) — T2/3
(d) — T4
(e) — T5
(f) — T6
(g) — T7
(h) — T8
(i) — T9
(j) — T10
(k) — T11
(l) — T12

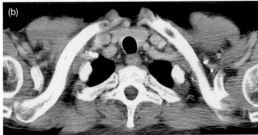

(b)

anterior jugular vein — thyroid
brachiocephalic artery — trachea
right subclavian vein — oesophagus
— left subclavian artery
head of humerus — lung apex
T1 — first rib
internal jugular vein — clavicle

5.6.1 (Continued)

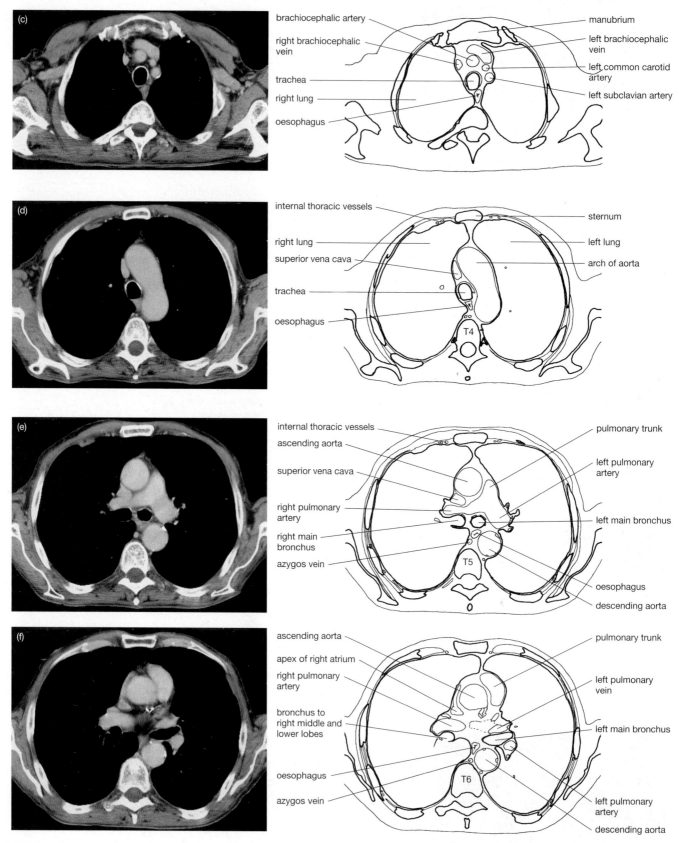

5.6.1 (Continued)

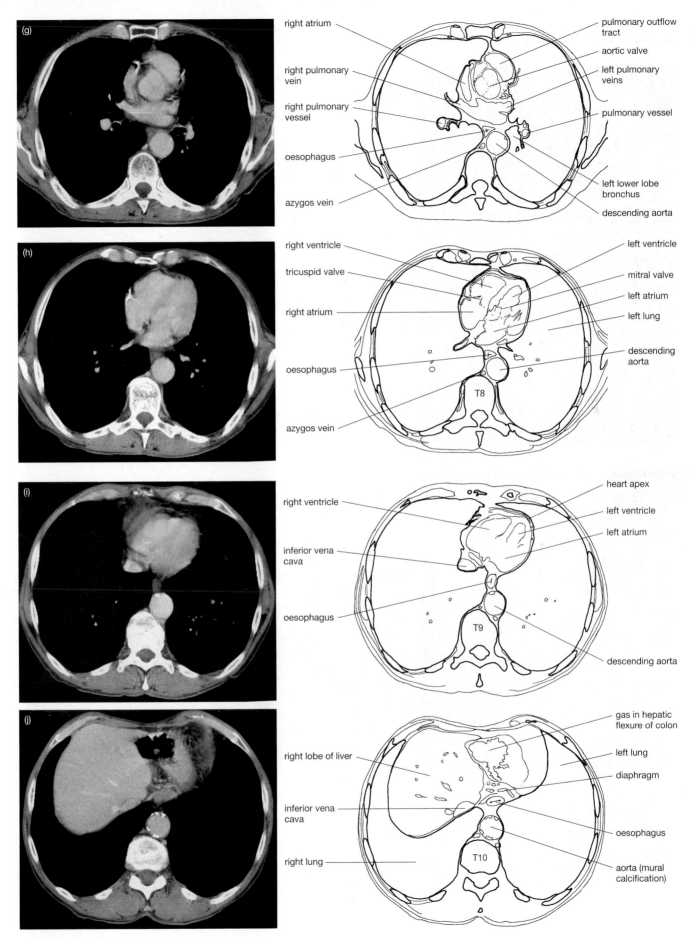

(g)

right atrium
right pulmonary vein
right pulmonary vessel
oesophagus
azygos vein

pulmonary outflow tract
aortic valve
left pulmonary veins
pulmonary vessel
left lower lobe bronchus
descending aorta

(h)

right ventricle
tricuspid valve
right atrium
oesophagus
azygos vein

left ventricle
mitral valve
left atrium
left lung
descending aorta

T8

(i)

right ventricle
inferior vena cava
oesophagus

heart apex
left ventricle
left atrium
descending aorta

T9

(j)

right lobe of liver
inferior vena cava
right lung

gas in hepatic flexure of colon
left lung
diaphragm
oesophagus
aorta (mural calcification)

T10

5.6.1 (Continued)

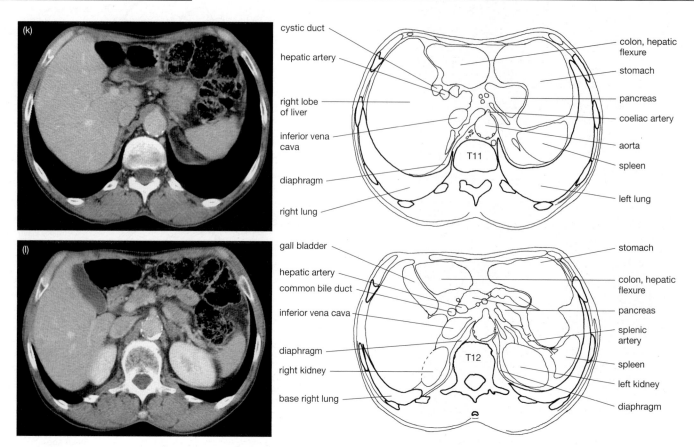

(k)

cystic duct
hepatic artery
right lobe of liver
inferior vena cava
diaphragm
right lung

colon, hepatic flexure
stomach
pancreas
coeliac artery
aorta
spleen
left lung

T11

(l)

gall bladder
hepatic artery
common bile duct
inferior vena cava
diaphragm
right kidney
base right lung

stomach
colon, hepatic flexure
pancreas
splenic artery
spleen
left kidney
diaphragm

T12

5.6.1 (a) Levels of the CT sections illustrated. (b)–(l) Computed tomograms and corresponding diagrams of cross-sections through different levels of the thorax. Some small patches of calcified atheroma are visible.

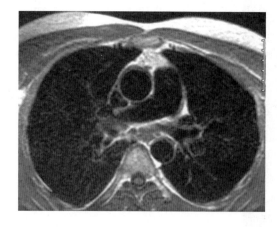

5.6.2 Axial MRI through the chest at T5/6 level. Compare with **5.6.1e**.

MRI **5.6.2** is an axial section (through the pulmonary trunk and right pulmonary artery) that can be compared with CT **5.6.1e**. MRI **5.5.7** is a sagittal section close to the midline. It shows the narrow anterior mediastinum, the heart and structures passing down through the superior and posterior mediastinum. See also axial MRI **5.4.17** which transects all four chambers of the heart, and coronal MRI **5.5.9** of the descending aorta.

Abdomen and pelvis: Introduction

Abdomen and pelvis: Introduction

The evolution of the primate upright stance was associated with many other adaptations. The chest became shorter and broader while the abdomen and pelvis also became foreshortened to form a somewhat small barrel-shaped cavity with protective walls of bone and muscle, situated between the thorax above and the perineum below. Within the abdominopelvic cavity lie organs and systems vital to the maintenance of the individual and the propagation of the species.

The **skin** of the trunk which surrounds the abdominopelvic cavity is marked by the umbilicus, a remnant of the point of attachment of the umbilical cord which links the fetus to the placenta and its mother. Abdominal and pubic hair is sexually dimorphic and dependent on androgens.

Beneath the skin of the back lies the spine with its associated **muscles**. These are surrounded by a strong lumbar fascia which, at its lateral margin gives attachment to three sheets of muscle which make the lateral and anterior walls of the abdominal cavity. Their flattened tendons (aponeuroses) ensheath the pair of vertical muscles of the anterior abdominal wall and then fuse in the anterior midline. With the diaphragm above and the muscles of the pelvic floor below, the abdominal wall muscles support the enclosed viscera. Their voluntary or reflex contraction not only protects the viscera but also raises the pressure in the abdominopelvic cavity, to aid forced expiration, coughing, defaecation and childbirth.

Lining the muscles of the abdominal walls, and covering many organs lying on its posterior aspect, is a layer of parietal serous **peritoneum**, which is continuous with a visceral layer covering much of the gut and other organs. This forms a closed sac, the peritoneal cavity, within the abdominopelvic cavity. Like the pleural cavity, the abdominal cavity is lubricated by a small amount of serous fluid, which allows the viscera to move on one another and on the abdominal walls. Double layers of peritoneum—mesenteries—attach the gut and its derivatives to the posterior abdominal wall; between their layers blood vessels, nerves, and lymphatics gain access to the gut. The mesenteries are often fan-shaped, allowing a long length of gut to be attached to a relatively short anchorage.

Within the protective walls of the cavity are a number of vital systems:

- The tubular **alimentary system** propels ingested food from the oesophagus, through the stomach and intestines to the anal canal, where unwanted remnants are excreted. With its derivatives, the liver, biliary system and pancreas, it is responsible for digesting food and absorbing the nutrients thus formed across the gut walls into the capillaries formed from its profuse blood supply. The **hepatic portal system** of veins carries the nutrient-rich blood to the **liver**, where the nutrients are metabolized and/or stored, controlling their concentrations in the systemic blood. The liver also produces bile, which contains pigment wastes and bile salts; the **biliary system** stores and transports bile to the gut where the salts are needed for emulsification of fats.
- The **urinary system** comprises the kidneys, which control the fluid and solute content of the blood, excreting wastes into the urine, which passes via the ureters into the urinary bladder where it is stored short-term until it is voided via the urethra.
- The **reproductive system** in the female comprises the ovaries, uterine tubes, uterus, and vagina; and in the male comprises the two testes, vas deferens, seminal vesicles, the prostate, and the penis. It is responsible not only for reproduction but also for much phenotypic sexual development. In males, production of healthy sperm requires a temperature lower than that in the abdominopelvic cavity. Therefore, during fetal life, the testes leave the abdominal cavity preceded by a sac of peritoneum and pass via a canal (the inguinal canal) through the anterior abdominal wall muscles in the groin, to reach the cooler scrotum.
- A number of abdominopelvic organs produce **hormones**. The endocrine pancreas, together with the intrinsic endocrine cells of the gut, regulates nutrient substrate availability and storage. The adrenal glands are essential for the body's response to stress: each comprises an inner medulla and a cortex which also regulates mineral balance. The gonads produce hormones which control phenotypic sexual development and the reproductive process. The liver and kidneys produce growth factors and also activate circulating hormone precursors.

All these systems derive arterial **blood** from the abdominal aorta and its branches and drain venous blood, either indirectly or directly, into the systemic inferior vena cava. Their rich **lymph** supply drains into the thoracic duct.

Somatic nerves from the lower thoracic, lumbar, and sacral segments of the spinal cord are segmentally distributed to the skin, muscles, and parietal peritoneum of the abdominal wall and perineum. The intra-abdominal organs and their visceral peritoneal coverings, however, receive an exclusively **autonomic nerve** supply from the sympathetic and parasympathetic systems. Their mutually complementary/antagonistic activity helps regulate the activity of the gut, the distribution of blood to the gut, systemic blood pressure, body temperature, and the gut, urinary, and reproductive reflexes.

Skeletal framework and subdivisions of the abdomen and pelvis

Skeletal framework and subdivisions of the abdomen and pelvis

The abdominal cavity is much larger than is apparent from looking at the anterior abdominal wall. The anterior wall consists only of soft tissues—skin, fascia, and muscle. The abdominal cavity, however, extends well above the anterior abdominal wall, so that its upper part lies within, and is protected by, the lower rib cage. The abdominal cavity also extends downward into the pelvic cavity, which is protected by a ring formed by the sacrum and pelvic bones. Posteriorly, the abdominal cavity is protected not only by the lumbar vertebrae but also by the lower thoracic spine and posterior parts of the lower ribs. In this way the soft organs of the abdomen and pelvis are substantially protected by bone and cartilage, and the vulnerable muscle wall is reduced to a minimum.

Skeletal framework

The skeletal framework (**6.1.1**) of the posterior wall of the abdomen and pelvis is formed by the spinal column (see Vol. 1, Chapter 7). The lower thoracic and **lumbar vertebrae** (**6.1.2**) form the posterior wall of the abdominal cavity; the **sacrum** forms the roof and posterior wall of the pelvic cavity.

Two **pelvic bones** articulate with the sacrum via the sacroiliac joints (see Vol. 1, p. 206). The pelvic bones are joined together in the midline by a (secondary) cartilaginous joint, the pubic symphysis. The bony pelvic girdle thus formed surrounds the lower part of the abdominal cavity and the pelvic cavity. Similarly, the upper part of the abdominal cavity is protected by the bone and cartilage of the lower thoracic vertebrae and ribs.

The pelvic girdle distributes body weight from the spine to the lower limbs and is adapted for bipedal locomotion. It also protects the outlets of the alimentary tract, urinary and genital systems.

The diaphragm separates the cavity of the abdomen from that of the thorax; muscles and fascia of the pelvic floor separate the pelvic cavity from the perineum.

The five **lumbar vertebrae** are heavily built. They have:

- **bodies**, which are wider transversely than anteroposteriorly;
- **transverse processes**, which are elongated and thin, with the exception of that of the fifth, which is conical and strong;
- **pedicles** from which protrude **articular processes**, which interlock to permit flexion and extension but preclude rotation of the lumbar spine (this interlocking starts with the lower articular process of T12 and continues downward, ending with the articulation between L5 and the sacrum);

- **laminae** from which protrude strong **spinous processes**, which project horizontally backward giving attachment to the extensor muscles of the spine.

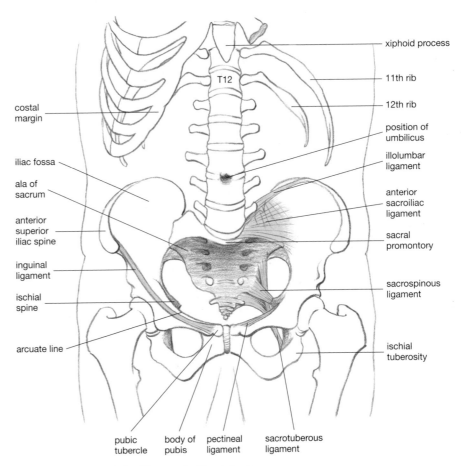

6.1.1 Boundaries and skeletal framework of the abdomen.

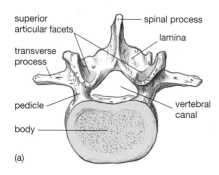

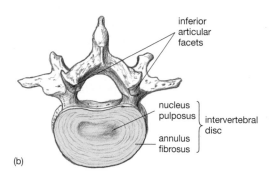

6.1.2 Lumbar vertebra viewed (a) from above, (b) from below; intervertebral disc.

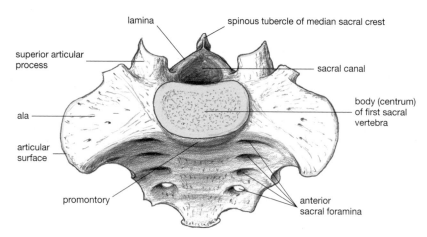

6.1.3 Sacrum viewed from above and in front.

The **sacrum** (6.1.3), composed usually of five fused vertebrae, is massive above, where it transmits forces between the pelvis and spine, but narrows rapidly below its articulation with the iliac bones. Identify:

- the **promontory**, which projects into the pelvic inlet in the anterior midline;
- the **ala** (wing), with its **articular surface** on both sides;
- anterior sacral foramina, which transmit the anterior primary rami of sacral spinal nerves as they enter the pelvis; the posterior primary rami pass through posterior foramina to supply extensor muscles and skin.

Strong **sacrotuberous** and **sacrospinous ligaments**, together with the ligaments around the sacroiliac joint, all stabilize the wedge-shaped sacrum between the two iliac bones of the pelvis and help resist the effect of body weight to push the sacrum downward and forward.

The **coccyx** consists of four rudimentary vertebrae, of which the first is the largest.

The anterior and lateral abdominal wall is bounded superiorly by the **xiphoid process, costal margin, eleventh** and **twelfth ribs** and inferiorly by parts of the **pelvis** (iliac crests and upper parts of the pubis).

The **pelvic bone** (6.1.4) is the bony girdle of the lower limbs. Unlike the upper limb girdle, which is adapted to give mobility to the limbs, the pelvic girdle is adapted for stability and weight-bearing. The female pelvis is also adapted for childbirth (p. 193).

It comprises three fused bones—the **ischium, ilium,** and **pubis**—and articulates on each side with the femur at the acetabulum. The three pelvic bones all have named prominences which serve as landmarks:

- **Ilium:**
  - **articular facet** for sacrum, and a roughened area behind the articular facet where the interosseous sacroiliac ligament is attached;
  - blade of the ilium, the concavity of which forms the **iliac fossa**, surmounted by the **iliac crest** and its **tubercle** (see Vol. 1, p. 130);
  - **anterior superior iliac spine**, anterior inferior spine, posterior superior, and inferior iliac spines;
  - the iliac component of the acetabulum;
  - stress-bearing thickened struts passing (1) from the acetabulum to the sacroiliac joint and (2) directly upward from the acetabulum.
- **Ischium:**
  - **ischial tuberosity**
  - **ischial spine**
  - ischial component of the acetabulum
  - ramus of the ischium which fuses with the inferior ramus of the pubis.
- **Pubis:**
  - **body of pubis**
  - pubic crest
  - **pubic tubercle**
  - pectineal line and pectineal surface
  - superior ramus of pubis
  - inferior ramus of pubis

- **vertebral foramina**, which are triangular in shape and larger than those of thoracic vertebrae.

The vertebrae and intervertebral discs, which are wider anteriorly than posteriorly, together form a **lumbar lordosis**.

The lumbosacral junction slopes downward and forward at about 35° to the horizontal. The iliolumbar ligaments, which link the transverse processes of L5 to the iliac crests, prevent the 5th lumbar vertebra from slipping forward on the sacrum.

Qu. 1A *What would be the consequences of a developmental anomaly (spondylolisthesis) in which the lower articular facets of L5 are not fused by bone to the centrum?*

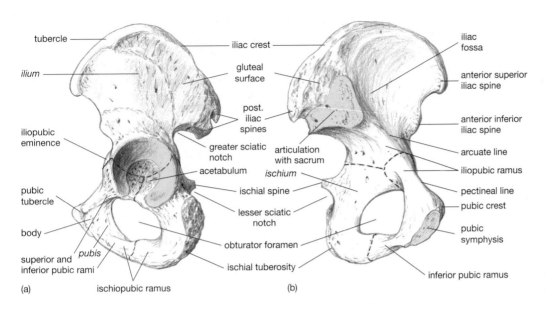

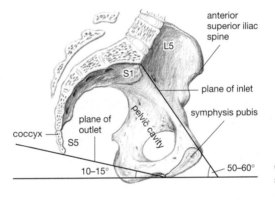

6.1.4 Pelvic bone: (a) outer and (b) inner aspects.

6.1.5 Sagittal section through pelvis showing planes of inlet and outlet.

– pubic component of the acetabulum

– articular surface for pubic symphysis.

The **acetabulum** has contributions from all three bones; the obturator foramen is bounded by the pubis and ischium; the greater sciatic notch lies above the ischial spine and is bounded by ischium and ilium; the lesser sciatic notch lies below the ischial spine.

The '**inlet**' **of the pelvis** (pelvic brim) is bounded by the promontory of the sacrum, and the **arcuate line**, the thickened part of the ilium running from the sacroiliac articulation to the superior ramus and body of the pubis and pubic symphysis. The '**outlet**' **of the pelvis** is bounded by the lower border of the pubic symphysis, inferior pubic and ischial ramus, ischial tuberosity, and coccyx.

The cavity of the bony pelvis is defined as that part of the abdominal cavity below the plane of the pelvic inlet.

It is important to understand the *orientation* of the pelvis (**6.1.5**). When standing erect the orientation of the pelvis is such that the anterior superior iliac spine(s) and the upper margin of the pubic symphysis/pubic tubercles lie in the same vertical coronal plane. The plane of the inlet of the pelvis therefore inclines forward at 50–60° to the horizontal; the plane of the outlet (from the tip of the coccyx to the under surface of the pubic symphysis) also inclines forward at about 15° to the horizontal in the midline.

Living anatomy

Identify the bony landmarks, define the subdivisions of the anterior abdominal wall, investigate the actions of the anterior abdominal wall muscles as a group, and begin to discover what physical examination can reveal about the abdominal organs.

Inspection

On yourself or, preferably, on a subject wearing trunks or a bikini, examine the **anterior abdominal wall** (**6.1.6**). Define the positions of the umbilicus, costal margin and xiphoid process, iliac crests ending in the anterior superior iliac spines, and the position of the midline pubic symphysis. When lying supine the abdominal wall is normally flat, though the amount of fat is very variable.

On either side of the midline is a vertical ridge of muscle (rectus abdominis) in which there are usually three horizontal tendinous intersections above the umbilicus (a remnant of embryonic segmentation). The muscle becomes more obvious on attempting to sit up. The lateral border of the muscle is marked by a **semilunar line**, which extends from the pubic tubercle upward to cross the costal margin at the tip of the ninth costal cartilage in the mid-clavicular line. On the right, this intersection marks the position of the gall bladder.

The **inguinal region** (groin) is the area at the junction of the anterior abdominal wall and lower limb.

A crescentic skin crease—the **groin** (**inguinal**) **crease**—runs from the anterior superior iliac spine downward and medially towards the midline, clearly demarcating the abdomen from the thigh. Immediately above the symphysis pubis is an eminence more marked in women than in men, the **mons pubis**.

Qu. 1B *What are the sex differences in the distribution of abdominal wall hair?*

Qu. 1C *What abnormalities might alter this distribution?*

On the back (**6.1.7**), a midline furrow overlies the concave lumbar region of the spine (lumbar lordosis) which is flanked by columns of spinal extensor muscle (erector spinae). Above the lumbar region the furrow becomes convex over the thoracic spine. Below the furrow is a flat triangular area of skin overlying the posterior aspect of the sacrum. At each lateral angle of this sacral triangle is usually a dimple marking the position of the posterior superior iliac spine. Passing upward and forward from the dimples, the curved iliac crests are clearly visible. At the inferior angle of the sacral triangle lies the coccyx, which curves forward deep within the natal cleft between the buttocks.

Compare the female torso (**6.1.7**) with that of the male (**6.1.6**): the wide shoulders and narrow hips of the male contrast with the more slender waist, wider hips and fatty buttocks of the female. These differences are secondary sex characteristics which become more apparent at puberty due to an increased output of sex steroids.

The anterior abdominal wall moves gently with respiration, and cardiac/arterial pulsations may be seen. It is normally impossible to see the outlines of any viscera or intestinal peristalsis.

For purposes of description, the anterior abdominal wall is divided by anatomists into nine regions (**6.1.8**). These are defined by two **lateral lines** running vertically from the mid point between the anterior superior iliac spine and the pubic symphysis; a **transpyloric plane** running horizontally at a level midway between the suprasternal notch and pubic symphysis (about midway between the umbilicus and the lower end of the sternum); and a **transtubercular plane** between the tubercles at the apices of the iliac crests. Both the transpyloric plane and the lateral line intersect the tip of the ninth costal cartilage (the surface marking of the tip of the gall bladder). The transpyloric plane is at the level of the lower border of the first lumbar vertebra; the transtubercular plane at the upper border of the fifth lumbar vertebra.

The nine regions thus defined (**6.1.8a**) are: right and left **hypochondrium** deep to the ribs separated by a midline **epigastrium**; a central **umbilical** region flanked by two **lumbar** regions; and a **suprapubic** region (hypogastrium), with two **iliac fossae**.

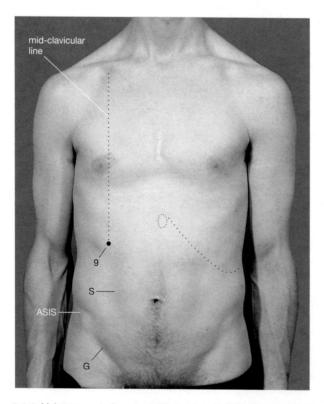

6.1.6 Male torso, anterior aspect. Rib margin and xiphisternum indicated on left. The mid-clavicular line intersects the tip of the ninth costal cartilage (9). Anterior superior iliac spine (ASIS); semilunar line marking lateral border of rectus abdominis (S); groin crease (G).

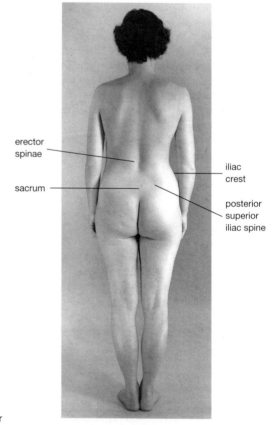

6.1.7 Posterior aspect of female torso.

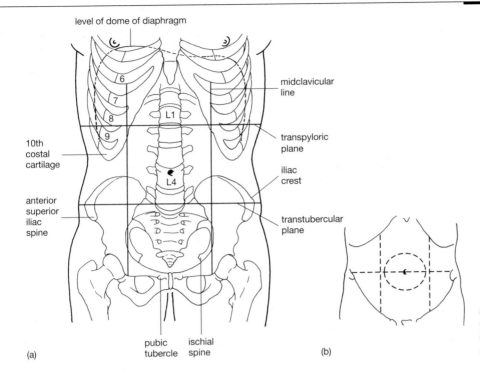

level of dome of diaphragm

midclavicular line

transpyloric plane

iliac crest

transtubercular plane

10th costal cartilage

anterior superior iliac spine

pubic tubercle   ischial spine

(a)

(b)

6.1.8 Diagram of abdomen and its skeletal framework (a) with the 'lines' used to divide the abdomen into regions. (b) Alternative nomenclature.

Surgeons may use a different nomenclature, dividing the abdomen horizontally through the umbilicus into an upper part comprising an epigastrium with right and left hypochondrium and a lower part comprising a hypogastrium with right and left iliac fossae, together with a central umbilical region (**6.1.8b**).

Palpation

Feel the bony margins of the anterior abdominal wall (**6.1.1, 6.1.8, 6.1.9**). Locate the xiphoid process and costal margin, noting that the tenth rib is the lowest in the mid-axillary line and the eleventh and twelfth ribs are difficult to feel. Palpate the spinous processes of the lumbar vertebrae. At the base of the sacrum the coccyx can also be felt; it can be injured if a person falls in the sitting position. Trace the iliac crest forward to the anterior superior iliac spine, noting the tubercle of the crest.

Medial to the anterior superior iliac spine the pelvic bone cannot be followed because of overlying muscles. In the midline, the pubic symphysis can be felt and the pubic crest followed laterally for a centimetre or so. The pubic tubercle lies on its anterior surface. To palpate the pubic tubercle place your hand flat on the upper aspect of the thigh with the pad of your middle finger lying on the crease above the mons pubis just lateral to the pubic symphysis (**6.1.9**). It is here that, in the male, the spermatic cord (p. 184) passes over the pubic tubercle, on its route between the scrotum and the abdomen.

Between the pubic tubercle and the anterior superior iliac spine stretches the **inguinal ligament**. This marks the lowest extent of the sheets of muscles that form the anterior abdominal wall laterally. More medially, rectus abdominis is attached to the pubic symphysis and crest.

Qu. 1D *By palpation, determine the actions of the anterior abdominal wall muscles. Ascertain when, and how strongly they contract:*
(a) *During a deep inspiration–expiration cycle.*
(b) *During a cycle of sitting up from the supine position and returning to it.*
(c) *When lifting a moderately heavy object from the floor by extension of the spine. Note the initial reflex inspiration before lifting starts. (Note, heavy objects should always be lifted with an upright spine and the knees flexed.)*
(d) *During simulated abdominal straining, during coughing, or as if constipated.*

Qu. 1E *What can you conclude about the non-postural actions of the anterior abdominal wall muscles as a group?*

Reflex contraction of the abdominal wall muscles in each of the four quadrants can be demonstrated by stroking the skin lightly outward from the umbilicus with a blunt-ended object (p. 230).

Qu. 1F *In what peripheral nerves does the reflex arc travel?*

Qu. 1G *What is the function of the reflex?*

Qu. 1H *If the reflexes were absent, what could you conclude?*

Qu. 1I *If the abdominal wall displays a board-like rigidity, what might be the cause?*

When attempting to palpate an abdomen, first sit so that your forearm and examining hand are held

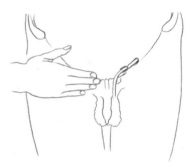

6.1.9 Palpation of pubic tubercle.

horizontally, with your eyes about 50 cm above your hand. This is the ideal position for viewing shadows formed by any lumps within the abdomen. To palpate any abdominal contents the abdominal wall muscles must be relaxed. The subject should therefore be lying comfortably, with knees bent if necessary, and your hand must be warm!

In the normal abdomen it is quite difficult to feel the outline of any viscus distinctly. The most readily palpable organ is the lower end of the **descending colon** in the left iliac fossa, because it is filled with relatively firm faeces. It may also be possible to feel the soft edge of a liver margin descending below the costal margin during inspiration; and the lower pole of a kidney may be palpable, especially if the kidneys are pushed forward by a hand supporting the loin. Most organs can be felt only if they are altered by disease or distended by the accumulation of their contents.

You can, however, usually feel the pulsations of the **abdominal aorta** by placing the palm of your hand on the anterior abdominal wall and pressing down gently with the pad of your fingers in the midline just below the umbilicus. The pulsations of the **femoral artery** can be felt as it passes under the inguinal ligament and into the leg (**6.1.9**) at a point midway between the anterior superior iliac spine and the pubic symphysis (mid-inguinal point).

During clinical examination, a lubricated, gloved finger is used to examine the anal and rectal canal, and the vagina and cervix (p. 197).

Percussion

Percussion of the abdomen will reveal solid or fluid-filled zones, such as the liver or a distended bladder, as 'dull' areas, and gas-containing viscera, such as the stomach and large bowel, as 'resonant' areas (**6.1.10**). Percuss and mark out the outline of the liver, especially its lower border beneath the right costal margin. Locate the resonant area in the left hypochondrium produced by the gas bubble in the stomach, and note any other resonant areas.

Auscultation

Place a stethoscope over the various regions of the anterior abdominal wall. Normally one can expect to hear 'bowel sounds' created by peristaltic movements of gas-containing contents of the bowel.

Imaging

Study the radiographs of the abdomen (**6.1.10**) and pelvis (**6.1.11**). Such 'straight' radiographs of the abdomen are usually taken antero-posteriorly.

Identify the outlines of the skeletal framework of the abdomen and pelvis. In addition to the bones, the lateral margins of psoas major and the lower border of the kidneys can often be distinguished (see **3.3**). Most other organs are seen only indistinctly or not at all on a plain film (though the liver margin and spleen may be visible) and special techniques are needed to reveal them. The bubble of gas in the stomach and the gas swallowed or formed by bacterial action in the colon are usually visible. Abnormally, gas may be present in the small intestine or may escape through a perforation in the bowel to enter the peritoneal cavity.

Abnormal calcification may be seen as stones in the biliary or urinary tracts, in the walls of diseased (atherosclerotic) arteries, in veins containing clotted blood (phleboliths), and in previously infected lymph nodes.

Qu. 1J *On radiographs 6.1.10 and 6.1.11 identify the structures marked A–D. What prominence is found at E?*

Questions and answers

Qu. 1A *What would be the consequences of a developmental anomaly (spondylolisthesis) in which the lower articular facets of L5 are not fused by bone to the centrum?*

Answer The body of L5 can slip forward and downward on the sloping upper surface of the sacrum,

6.1.10 Antero-posterior radiograph of the normal abdomen.

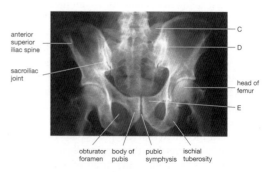

6.1.11 Radiograph of the normal pelvis.

carrying the upper spinal column and cord with it. This will cause chronic backache, and can put tension on, or compress, the cauda equina of nerve roots within the spinal canal.

Qu. 1B *What are the sex differences in the distribution of abdominal wall hair?*

Answer Sexually dimorphic hair begins to develop at puberty. In men, coarse hair extends up from the pubic region to the umbilicus (and may cover more of the abdominal wall); in women, pubic hair stops at the upper border of the mons pubis.

Qu. 1C *What abnormalities might alter this distribution?*

Answer Abnormally increased androgen production in a woman results in male-type hair distribution. Absence of androgen-dependent hair may indicate an absence of androgen receptors in a genetic male (androgen insensitivity).

Qu. 1D *By palpation, determine the actions of the anterior abdominal wall muscles. Ascertain when, and how strongly they contract:*
(a) During a deep inspiration–expiration cycle.
(b) During a cycle of sitting up from the supine position and returning to it.
(c) When lifting a moderately heavy object from the floor by extension of the spine. Note the initial reflex inspiration before lifting starts. (Note, heavy objects should always be lifted with an upright spine and the knees flexed.)
(d) During simulated abdominal straining, during coughing, or as if constipated.

Answer Anterior abdominal wall muscles contract:
(a) During expiration, especially at the end of forced expiration.
(b) Rectus abdominis, in particular, contracts throughout a sitting/lying cycle, except in the upright position, at which point balance requires minimal effort; its contraction provides power to lift the trunk, and then resists gravity during the controlled return to the supine position.
(c) Before a lifting effort starts, an inspiration occurs and the glottis of the larynx is closed to prevent exhalation and to maintain intrathoracic volume. At the same time, the abdominal muscles contract

to maintain high intra-abdominal pressure during spinal extension.
(d) The muscles contract strongly during straining (e.g. childbirth) and coughing.

Qu. 1E *What can you conclude about the non-postural actions of the anterior abdominal wall muscles as a group?*

Answer The muscles provide support and protection for the anterior abdominal wall and compress the abdominal contents. If the diaphragm also contracts, abdominal pressure is raised, as in vomiting. Rectus abdominis also acts as flexor of the lumbar spine; the more lateral muscles aid oblique flexion of the trunk.

Qu. 1F *In what peripheral nerves does the reflex arc travel?*

Answer The afferent and efferent limbs of the abdominal reflex pass in the lower intercostal nerves (T6–T12).

Qu. 1G *What is the function of the reflex?*

Answer The reflex is protective. Mild unexpected stimuli cause contraction of the abdominal wall, which protects the abdominal viscera.

Qu. 1H *If the reflexes were absent, what could you conclude?*

Answer Absence of abdominal reflexes indicates non-functioning of the thoracic afferent (sensory) or efferent (motor) neurons, or of the spinal interneurons which complete the reflex arc.

Qu. 1I *If the abdominal wall displays a board-like rigidity, what might be the cause?*

Answer Irritation of the parietal peritoneum of the anterior abdominal wall causes powerful reflex contraction of anterior wall musculature. Blood, acid, and inflammatory loci are all powerful irritants of parietal peritoneum.

Qu. 1J *On radiographs 6.1.10 and 6.1.11 identify the structures marked A–D. What prominence is found at E?*

Answer A, spinous process of L3; B, twelfth rib; C, transverse process of L5; D, ala of sacrum; E, pubic tubercle.

Boundaries of the abdomen and pelvis I: Anterior abdominal wall and inguinal region

Boundaries of the abdomen and pelvis I: Anterior abdominal wall and inguinal region

The anterior abdominal wall muscles protect the abdominal viscera and raise intra-abdominal pressure. They consist, on each side, of three flat sheets of muscle and aponeuroses which extend around the sides and front of the abdominal cavity, and one vertically placed muscle. These are covered by skin and layers of fat-storing and fibrous superficial fascia, and are lined internally with peritoneum. All these are supplied primarily by the lower six intercostal nerves and vessels. In males before birth, the testes, guided by the gubernaculum, migrate through the wall just above the inguinal ligament to reach the scrotum. This migration forms a fibromuscular (inguinal) canal, the internal and external openings of which form areas of potential weakness through which hernias of abdominal contents can occur. The canal is also present in women, but is much smaller, and contains the round ligament.

Development

The **fascia** and **muscles** of the antero-lateral abdominal walls develop in the lateral plate mesoderm of the body wall, which folds laterally around the developing abdominal cavity and viscera. The lateral folding, together with the craniocaudal folding (which carries the septum transversum and heart beneath the developing foregut), causes mesoderm with its covering ectoderm to form a complete ventro-lateral body wall around the embryo, except in the region of the developing umbilical cord. The umbilical cord provides a space into which the developing gut temporarily herniates; but this becomes progressively smaller as the definitive umbilical cord is formed. As in the thoracic region, the mesoderm forms three layers of muscle. Nerves and vessels grow into the developing body wall from the aorta and spinal cord, forming a neurovascular plane (p. 116) between the middle and innermost layers of muscle.

The **testis** develops on the posterior wall of the developing abdominal cavity (p. 182). However, around the time of birth, the testis and its duct system (epididymis and vas deferens), with their associated blood and lymph vessels and nerves (known collectively as the spermatic cord), migrate through the anterior abdominal wall to reach the **scrotum**, a pouch of skin which forms in the anterior part of the perineum.

Qu. 2A *For what physiological process is descent of the testis into the scrotum essential?*

In their migration through the lower part of the anterior abdominal wall, the migrating testis and cord form a fibromuscular channel in the lower extremity of the anterior abdominal wall—the **inguinal canal**—and derive a covering from the layers of fascia

and muscles through which they pass. The testis is preceded in its descent by a diverticulum of the peritoneal cavity—the **processus vaginalis** (see **6.2.1a**)—which is guided by a band of specialized connective tissue—the **gubernaculum**—attached between the testis and scrotum. Normally the proximal part of the peritoneal diverticulum closes off by fibrosis, preventing herniation of intra-peritoneal contents. Its distal part remains as a bursa (**tunica vaginalis**) which partly surrounds the testis (see **6.2.1b**).

In the female, the gubernaculum extends from the ovary to the labia majora (p. 209). However, the ovary descends only as far as the pelvic cavity, so that the inguinal canal is much smaller and contains only the derivative of the distal part of the gubernaculum, the **round ligament** of the uterus.

Subcutaneous tissue of the anterior abdominal wall

Immediately beneath the skin is a layer of (white) **subcutaneous fat**, which is continuous with that in the thorax and extends over the inguinal region and into the thighs. The fat is very variable in extent and, because fat has relatively few blood vessels, thick layers of fat may heal poorly after abdominal incisions. In women it is continuous with the fat of the mons pubis and labia majora (p. 209). In men the subcutaneous tissue of the scrotum is usually devoid of fat, but it does contain smooth muscle—**dartos**—which, on contraction, wrinkles the scrotal skin and moves the testis closer to the body wall.

The deepest layer of superficial fascia is a **membranous layer** (Scarpa's fascia) which contains many collagen and elastic fibres. This is most marked below the level of the umbilicus, where it helps support the

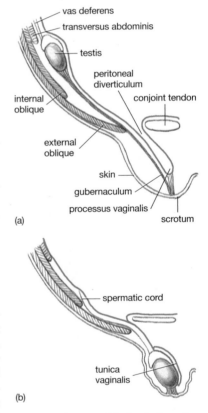

(a)

(b)

6.2.1 Development of the inguinal canal: descent of the testis preceded by the vaginal process.

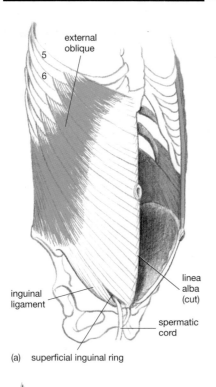

external
oblique

5
6

inguinal
ligament

spermatic
cord

linea
alba
(cut)

(a)   superficial inguinal ring

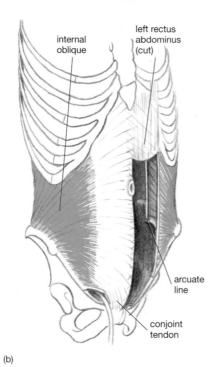

internal
oblique

left rectus
abdominus
(cut)

arcuate
line

conjoint
tendon

(b)

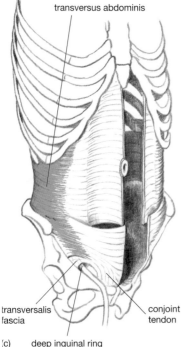

transversus abdominis

transversalis
fascia

conjoint
tendon

(c)   deep inguinal ring

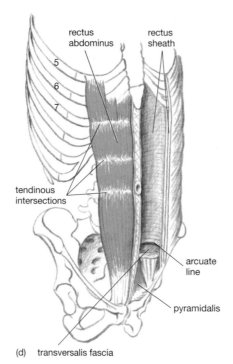

rectus
abdominus

rectus
sheath

5
6
7

tendinous
intersections

arcuate
line

pyramidalis

(d)   transversalis fascia

6.2.2 Muscles of the anterior abdominal
wall: (a) external oblique; (b) internal
oblique; (c) transversus abdominis;
(d) rectus abdominis; (e) section through
anterior abdominal wall. The aponeuroses
of the abdominal wall muscles which form
the rectus sheath interdigitate to form the
midline linea alba.

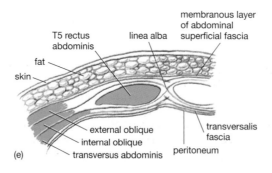

T5 rectus
abdominis

fat

skin

linea alba

membranous layer
of abdominal
superficial fascia

external oblique
internal oblique
transversus abdominis

transversalis
fascia

peritoneum

(e)

abdominal contents. It extends around the shaft of the
penis and beneath dartos to form a fascial layer around
the scrotum (see **6.10.6–6.10.8**). The membranous
layer of superficial fascia is firmly adherent to:
- the body of the pubis and inferior pubic rami;
- the posterior edge of the perineal membrane (p. 206);
- the deep fascia (fascia lata) of the thigh about 1 cm
  below the inguinal ligament.

The clinical significance of these attachments is that
they prevent any abnormal accumulation of fluid
beneath the membranous layer from passing either
into the subcutaneous tissue of the legs or into the anal
region of the perineum (p. 206).

Muscles of the anterior abdominal wall

The muscles of the anterior abdominal wall comprise
three flat sheets of muscle lying one beneath the other
on each side:
- external oblique
- internal oblique
- transversus abdominis,
and a vertically oriented pair of muscles, the fibrous
sheaths of which are joined in the midline:
- rectus abdominis.
   These muscles all act:
- to support and compress the abdominal contents;
- if the airway is open, to force the diaphragm upward
  (forced expiration);
- if the airway is closed, to raise the intra-abdominal
  pressure (e.g. in abdominal straining, vomiting,
  coughing, childbirth).
Rectus abdominis is also a powerful flexor of the spine;
the oblique muscles aid rotation (twisting) of the
trunk, which occurs largely in the thoracic spine.

The abdominal wall muscles and the overlying skin
and fascia are all supplied by the lower intercostal
nerves (T6–T12; p. 117).

External oblique (**6.2.2a**) is the most superficial. Its
muscular fibres arise from the outer surface of the
lower eight ribs and pass downward and medially.
Posteriorly the fibres form a free (unattached) border
and insert into the iliac crest forward to the anterior
superior iliac spine. Here the muscle fibres give way to
a broad **aponeurosis** which continues to the midline,
covering rectus abdominis, and interlacing with the
aponeurosis of the other side to form the **linea alba**
(white line), which extends between the xiphoid
process and the pubic symphysis.

Between the anterior superior iliac spine and the
pubic tubercle, the aponeurosis is thickened and
folded back on itself to form an upward-directed gutter, the **inguinal ligament**. Immediately above and lateral to the pubic tubercle the aponeurotic fibres
diverge to form an apparent triangular gap, the **superficial inguinal ring** (**6.2.2a, 6.2.3**), the medial and lateral margins of which are thickened (crura). As the
spermatic cord and testis pass through the superficial

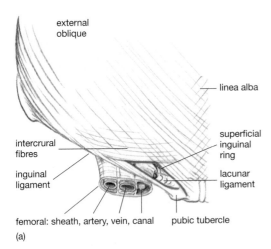

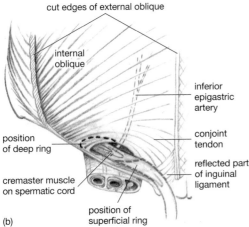

6.2.3 (a) Position of superficial inguinal ring relative to the femoral sheath and its contents. Note the sharp edge of the lacuna ligament in relation to the femoral canal, which is continuous with the abdominal (extraperitoneal) cavity. (b) External oblique has been largely removed to show the floor and roof of the inguinal canal and the spermatic cord.

inguinal ring they acquire a thin covering from external oblique—the **external spermatic fascia**.

The superficial inguinal ring can be felt by invaginating the skin of the scrotum with the little finger and following the cord proximally.

Fibres reflected from the medial end of the inguinal ligament extend backward, upward and laterally from the pubic tubercle to form a **lacunar ligament**. This attaches to and extends along the medial part of the pectineal line as the **pectineal ligament**. The lacunar ligament is triangular in shape and its sharp, free lateral edge forms the medial boundary of the femoral canal (**6.2.3**).

Internal oblique (**6.2.2b**) lies deep to external oblique. Its muscular fibres arise from the lateral two-thirds of the inguinal ligament, the anterior part of the iliac crest and the lumbar fascia. Its posterior fibres pass upward and medially to insert into the costal margin; more medial fibres form an aponeurosis that interlaces with that of the opposite side in the linea alba; the fibres arising from the inguinal ligament arch

medially and downward, passing over and then behind the spermatic cord (or round ligament) and, becoming aponeurotic, insert with similar fibres from transversus abdominis into the crest of the pubis. In this way a **conjoint tendon** is formed which reinforces the posterior wall of the superficial inguinal ring. The upper aponeurotic fibres attaching to the linea alba divide to ensheath rectus abdominis, the lower fibres pass entirely superficial to that muscle.

Internal oblique contributes a thin covering to the testis and cord, the **cremaster** muscle and fascia. Muscle fibres of cremaster loop around the cord and are attached to the **tunica vaginalis**. They are supplied by the genital branch of the genito-femoral nerve (L2; p. 126). The **cremasteric reflex** (p. 230) by which the testis is lifted upward can be elicited by stroking the inner aspect of the thigh (L2 innervation). It is more active in children than in adults and this can result in the misdiagnosis of undescended testes, particularly if the vaginal process has not completely closed off.

Transversus abdominis (**6.2.2c**) is the innermost of the three muscle sheets. Its fibres arise deep to internal oblique from the lateral half of the inguinal ligament, the iliac crest, the lumbar fascia, and from the inner aspect of the lower six costal cartilages, interdigitating with those of the diaphragm. Most of its fibres pass horizontally, become aponeurotic, and pass behind the upper part of rectus abdominis and in front of its lower part. The fibres originating from the inguinal ligament arch downward and medially behind the spermatic cord, joining with fibres from internal oblique to form the conjoint tendon.

The **transversalis fascia** is a thin sheet of connective tissue which lies between transversus and the underlying peritoneum. It is most prominent in the inguinal region, and is continued over the spermatic cord as the **internal spermatic fascia**.

Rectus abdominis (**6.2.2d**) arises from the pubic symphysis and crest and, becoming broader, passes upward to cross the costal margin and attach to the fifth, sixth, and seventh costal cartilages. It is enclosed in the **rectus sheath** formed by the aponeuroses of the other abdominal wall muscles. The posterior wall of the sheath is deficient in the lower quarter of the abdomen where all the aponeuroses pass superficial to the muscle.

Inguinal canal

The inguinal canal (**6.2.3**) is a passage through the lower anterior abdominal wall which transmits the spermatic cord in the male (the round ligament in the female) between the abdominal cavity and the scrotum (labia majora). It lies parallel with and just above the medial half of the inguinal ligament and, in the adult male, is about 4 cm long.. It extends from the **deep inguinal ring**, bounded laterally by the most medial fibres of transversus abdominis, to the **superficial inguinal ring**, the gap in the aponeurosis of external oblique, lying just above the pubic tubercle.

Hernias

Raised intra-abdominal pressure will tend to force abdominal contents out through any point of weakness. Such a protrusion is called a **hernia**. The contents of a hernial sac may be restricted to pieces of fatty omentum but loops of small bowel or other mobile organs can also enter.

Inguinal hernia

An inguinal hernia may be either:
- **indirect** (6.2.4a)—passing along the length of the inguinal canal, usually in young people, or
- **direct** (6.2.4b, 6.2.5)—often in older people protruding through a weakened conjoint tendon.
  Indirect inguinal hernias leave the abdominal cavity lateral to the inferior epigastric artery, whereas direct inguinal hernias bulge the weakened abdominal wall medial to the artery. Indirect hernias emerge through the superficial inguinal ring and extend into the scrotum. Direct hernias usually do not pass through the superficial ring or into the scrotum. Some hernias are 'reducible' i.e. they can be returned to the abdominal cavity by manual manipulation or by lying supine; others cannot and are said to be 'irreducible'. Any hernia that becomes trapped can have the blood supply to its contents impaired ('strangulation'). If this happens, the contents can die (necrose). If loops of bowel become strangulated, their walls will break down, and may perforate.

Femoral hernia

A nearby area of weakness is the **femoral canal** (Vol. 1, p. 179). This short, tapering space lies beneath the medial part of the inguinal ligament, between the femoral vein (laterally) and lacunar ligament (medially). It permits expansion of the femoral vein and transmits lymphatics from the leg to abdominal lymph nodes. Through this canal a **femoral hernia (6.2.4c)** may protrude. The greater width of the female pelvis coupled with the small size of the round ligament means that, in women, femoral hernias are relatively more common than inguinal hernias, although both types of hernia occur more commonly in men.

Femoral hernias are usually much smaller than inguinal hernias, but the sharp margin of the lacunar ligament means that they are more commonly strangulated.

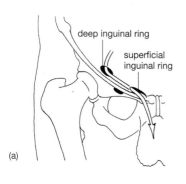

(a)

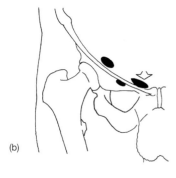

(b)

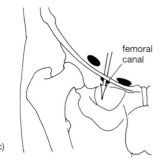

(c)

6.2.4 The paths taken by (a) an indirect inguinal hernia, (b) a direct inguinal hernia, and (c) a femoral hernia.

The inferior epigastric branch of the external iliac artery (see below), passes upward to form a medial border to the deep ring. The ilio-inguinal nerve (see below) runs with the cord in the medial part of the canal; it supplies the lower fibres of internal oblique in addition to the skin of the scrotum.

The walls of the inguinal canal are:
- **floor:** the internal aspect of the inguinal ligament;
- **anterior wall:** external oblique, reinforced laterally by internal oblique covering the deep inguinal ring;
- **roof:** arching fibres of transversus abdominis and internal oblique;
- **posterior wall:** transversalis fascia reinforced medially by the conjoint tendon lying behind the superficial inguinal ring.

The inguinal canal transmits, in the male, structures which form the spermatic cord. This contains the:
- vas deferens
- obliterated processus vaginalis
- testicular artery and plexus of veins
- artery to the vas deferens
- lymphatics draining testis, epididymis, and vas deferens,

successively covered by:
- internal spermatic fascia (from transversus abdominis and the underlying fascia)
- cremaster muscle and fascia (from internal oblique)
- external spermatic fascia (from external oblique),

and the ilio-inguinal nerve (supplying scrotal/labial skin).

The inguinal canal is an obvious point of weakness in the anterior abdominal wall, although its oblique passage through the muscle layers helps to compensate for this. It is weaker in the male than the female because of the larger size of the spermatic cord compared to the round ligament. This weakness is compounded if the (peritoneal) vaginal process remains partly or completely patent (open).

Qu. 2B *What other factors might be involved in the greater incidence of hernias in men than in women?*

Qu. 2C *How might you distinguish between a direct and an indirect inguinal hernia, provided that the hernia was reducible?*

Qu. 2D *Which bony landmark(s) would you use to distinguish a femoral from an inguinal hernia?*

Vessels and nerves of the anterior abdominal wall

The lower six **intercostal nerves** and their accompanying **arteries, veins,** and **lymphatics** (6.2.6, 6.2.7) pass deep to the costal margin to enter and then run obliquely medially around the anterior abdominal wall. They lie in the **neurovascular plane** between internal oblique and transversus abdominis and end by entering the rectus sheath.

Having supplied their thoracic segment they also supply:
- the muscles of the anterior abdominal wall with motor and sensory fibres;
- sensation to a segmental strip of skin over the side and front of the abdomen (by lateral and terminal cutaneous branches);

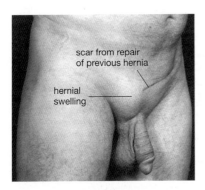

6.2.5 Photograph of direct inguinal hernia.

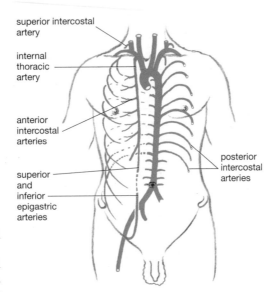

6.2.6 Arteries of the anterior abdominal wall.

- sensation to a corresponding area of parietal peritoneum.

It is helpful to remember that the sixth intercostal nerve (T6) supplies the region of the xiphoid process, T10 the region around the umbilicus, and T12 the region of the pubic symphysis.

Qu. 2E *How does this arrangement correspond to that of the intercostal nerve supply to the chest wall?*

The **subcostal nerve** (T12) passes in the neurovascular plane below the costal margin to supply a strip of body wall parallel to and just inferior to the eleventh intercostal nerve.

The **iliohypogastric** and **ilio-inguinal nerves** are both derived from the anterior primary ramus of L1. They follow a parallel course to the subcostal nerve. The iliohypogastric nerve ends by supplying skin over the pubic symphysis.

The **ilio-inguinal nerve** pierces the internal oblique to enter the inguinal canal and emerges through the superficial inguinal ring to enter the scrotum or labium majus. It supplies:

- the lowest muscular fibres of internal oblique and transversus, which are attached to the conjoint tendon;
- cremaster (derived from internal oblique);
- skin over the root of the penis and the scrotum (or mons pubis and labium majus) and an adjacent part of the medial aspect of the thigh;
- a corresponding strip of parietal peritoneum.

Qu. 2F *Why is it important for the surgeon carefully to preserve the ilio-inguinal nerve when repairing an inguinal hernia or making a gridiron incision for appendicitis (see below)?*

Qu. 2G *Why would the loss of a single sensory nerve root not lead to a strip of anaesthetic skin in the distribution of the nerve, whereas irritation of a single nerve root would give rise to a distinct band of pain?*

The lower six **posterior intercostal arteries** run with the corresponding nerves to supply the anterior abdominal wall. In the rectus sheath the supply is supplemented by the **superior** and **inferior epigastric arteries** (6.2.6). The superior epigastric artery is a terminal branch of the internal thoracic artery; the inferior is a branch of the external iliac artery given off just

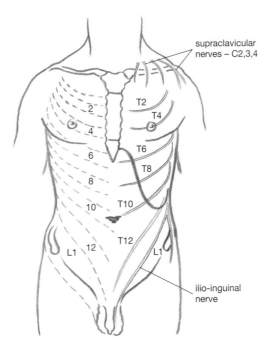

6.2.7 Nerves of the anterior abdominal wall. T7, 9, and 11 are omitted.

before that vessel passes beneath the inguinal ligament to form the femoral artery. The inferior epigastric artery passes upward, medial to the deep inguinal ring in the transversalis fascia, to reach the deep aspect of rectus abdominis. It passes up in its sheath to anastomose with the superior epigastric artery and the intercostal arteries.

Qu. 2H *In what major congenital malformation of the arterial tree would you expect this arterial anastomosis to enlarge?*

Superficial **veins** of the anterior abdominal wall drain up to the axilla or down to (leg) veins in the groin; deeper veins run with the arteries. Superficial **lymphatics** follow the veins to superficial axillary or inguinal lymph nodes; a few deep lymphatics follow the arteries.

At the umbilicus there is an anastomosis between veins of the abdominal wall and those in the falciform

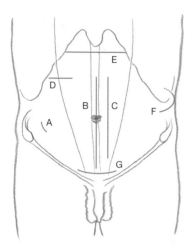

6.2.8 Incisions for entering the abdominal cavity.

Abdominal incisions

Incisions to open the abdominal cavity (**6.2.8**) are usually made through the anterior abdominal wall, which has to withstand considerable rises in intra-abdominal pressure. The principles of abdominal incisions are therefore to minimize the weakness that would result from transverse cuts across muscle fibres, and to minimize the cutting of nerves to reduce postoperative pain, while at the same time providing sufficient access for internal examination (laparotomy).

The **gridiron** incision (A) approach to an appendix involves a small incision in the right iliac fossa, centred on the surface marking of the root of the appendix (McBurney's point; p. 155). The three exposed layers of muscle and aponeurotic fibres can then be separated along their length without cutting, to expose the peritoneum and thereby the appendix.

If greater access to the abdominal cavity is needed, a **midline** incision (B) through the linea alba is often used. This avoids cutting muscles or nerves and has largely replaced the 'paramedian' incision (C) which involved vertical incisions through the rectus sheath (retracting the rectus muscle laterally to reach the posterior wall of the sheath).

A variety of **transverse** incisions are also used. These are designed to avoid cutting across dermatomes. Transverse incisions are made in the right hypochondrium to approach the gall bladder (D); beneath the rib margin in the midline (E; 'roof-top') to approach the liver, stomach, or pancreas; laterally in the loin to approach the kidney (F); and in the suprapubic region to approach the bladder (G).

Skin minimal tension lines

These lines of minimal tension (see **2.4**) are determined by the arrangement of collagen fibres in the dermis. They are important in surgery because incisions *along* the lines heal with minimal scarring; incisions *across* the lines can produce more prominent, heaped-up scars.

ligament of the liver. These can enlarge if there is high pressure in the hepatic portal system of veins (p. 163).

Questions and answers

Qu. 2A *For what physiological process is descent of the testis into the scrotum essential?*

Answer Spermatogenesis, which requires, in man, a temperature several degrees lower than that which obtains in the abdomen.

Qu. 2B *What other factors might be involved in the greater incidence of hernias in men than in women?*

Answer The manual work that many men do involves heavy lifting, with increased abdominal pressure; also, hypertrophy of the prostate (see p. 186) causes many older men to have to strain to pass urine.

Qu. 2C *How might you distinguish between a direct and an indirect inguinal hernia, provided that the hernia was reducible?*

Answer If the hernia has been reduced and the patient is asked to cough, the increase in abdominal pressure will force the hernia through its opening in the anterior abdominal wall. If this is a congenital hernia, it will thrust obliquely through the deep inguinal ring, following the path of the spermatic cord to the superficial inguinal ring. A direct hernia will thrust directly through the abdominal wall to emerge at the superficial inguinal ring.

Qu. 2D *Which bony landmark(s) would you use to distinguish a femoral from an inguinal hernia?*

Answer The pubic tubercle. An inguinal hernia lies above the tubercle, a femoral hernia lies below.

Qu. 2E *How does this arrangement correspond to that of the intercostal nerve supply to the chest wall?*

Answer The cutaneous nerves emerge as a lateral group (along the anterior axillary line) and an anterior group (through rectus abdominis) as in the thorax. Supply to the xiphisternum is from T6; the umbilicus, T10; and the pubic symphysis, T12.

Qu. 2F *Why is it important for the surgeon carefully to preserve the ilioinguinal nerve when repairing an inguinal hernia, or making a gridiron incision for appendicitis?*

Answer The ilioinguinal nerve must be preserved because it supplies the lowest fibres of external and internal oblique and transversus (and cremaster) in addition to the skin of the scrotum. If the nerve supply to is damaged, the conjoint tendon would be weakened and result in the formation of a direct hernia.

Qu. 2G *Why would the loss of a single sensory nerve root not lead to a strip of anaesthetic skin in the distribution of the nerve whereas irritation of a single nerve root would give rise to pain?*

Answer There is considerable overlap in the areas of skin supplied by sensory spinal nerves, so that loss of a single nerve would not be noticed. However, stimulation of a single nerve root causes pain in the distribution of the root.

Qu. 2H *In what major congenital malformation of the arterial tree would you expect this arterial anastomosis to enlarge?*

Answer Coarctation of the aorta.

Boundaries of the abdomen and pelvis II: Posterior abdominal and pelvic walls, vessels, and nerves

Boundaries of the abdomen and pelvis II: Posterior abdominal and pelvic walls, vessels, and nerves

This section considers the remaining boundaries (posterior wall and roof) of the abdominal cavity, and the walls and floor of the pelvic cavity. The posterior wall of the abdominal cavity is formed by the lumbar vertebrae, their associated muscles and fascia; its domed roof is formed by the diaphragm. The walls of the pelvic cavity are formed by the sacrum, pelvic bones, and their associated muscles and fascia; its floor is formed by the levator ani muscle and the fascia of the pelvic floor. On the posterior and lateral walls of the abdomen and pelvis are many major vessels and nerves. These are considered here because they supply the abdominal and pelvic organs, which are the subject of subsequent sections.

Posterior wall of the abdominal cavity

The posterior abdominal wall (**6.3.1**) is formed in the midline by the bodies of the **lumbar vertebrae** flanked on either side by the crura of the diaphragm and by **psoas** major, an elongated muscle which extends across the pelvic brim. Together these form a pronounced midline ridge which projects into the abdominal cavity.

Lateral to psoas is a deep **paravertebral gutter**, in the floor of which **quadratus lumborum** extends between the twelfth rib and the iliac crest. Above quadratus lumborum is the posterior part of the diaphragm, lateral to quadratus lumborum is transversus abdominis, while **iliacus** fills the iliac fossa below quadratus lumborum.

Psoas (major) (**6.3.1**) flexes the hip and the lumbar spine (Vol. 1, p. 141). It arises from the sides of the bodies of T12 and the lumbar vertebrae, from the intervertebral discs between them, and from the lumbar transverse processes. Its tendon passes beneath the inguinal ligament, crosses the anterior aspect of the hip joint, and is attached to the lesser trochanter of the femur. It is invested in a tight fascial sheath and is supplied segmentally by lumbar nerves.

Qu. 3A *If pus accumulated within the psoas sheath (e.g. from a tuberculous vertebra), where on the skin would it be likely to discharge ('point')?*

Iliacus (**6.3.1**) arises from the upper two-thirds of the iliac fossa and is also covered by a tough fascial sheet. It passes beneath the inguinal ligament lateral to psoas, inserting partly into the psoas tendon and partly into the femur just below the lesser tuberosity. It is supplied by the femoral nerve.

Qu. 3B *Which action of psoas is shared by iliacus and which is not?*

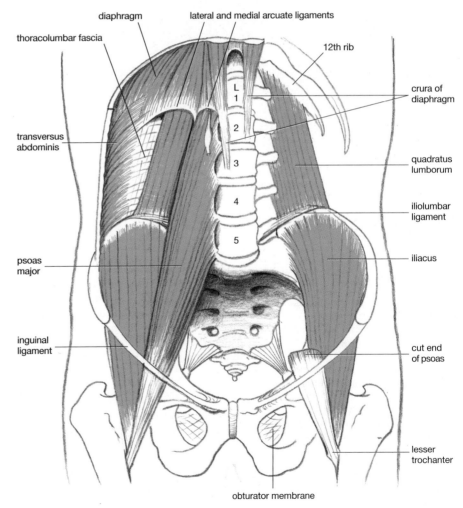

6.3.1 Muscles and bones of the posterior abdominal wall.

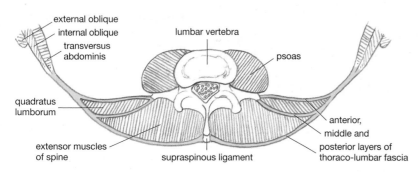

6.3.2 Thoraco-lumbar fascia.

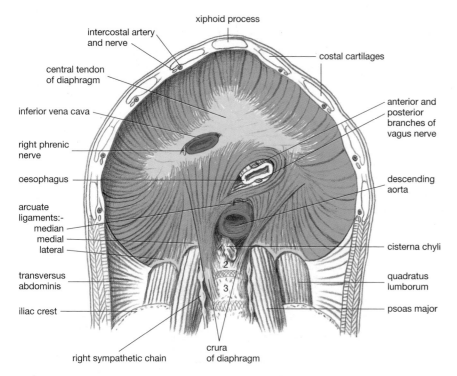

6.3.3 The diaphragm forming the roof of the abdomen.

Roof of the abdominal cavity

The **diaphragm** forms a thin but strong domed roof over the abdominal cavity (**6.3.3**; see **5.2.2**). It is covered by pleural and pericardial membranes on its thoracic surface and by peritoneum on most of its abdominal surface. Its function in inspiration has been considered. In expanding the thoracic cavity it also raises intra-abdominal pressure. Its crura help prevent reflux up the oesophagus.

On each side, it arises from:
- the inner aspect of the **xiphoid** process, **costal margin, eleventh** and **twelfth** ribs;
- a lateral arcuate ligament, which arches over quadratus lumborum to the transverse processes of L1, and a medial arcuate ligament which arches over psoas to the body of L2;

and, posteriorly, from
- two muscular **crura** attached to the antero-lateral surface of the **bodies of upper lumbar vertebrae** (right, L1, 2, and 3; left, L1 and 2) and united in the midline.

Some fibres of the right crus loop to the left around the lower end of the oesophagus; this muscular loop helps to maintain the continence of the gastro-oesophageal junction when intra-abdominal pressure is raised.

Qu. 3D *Which structures enter/leave the abdomen between the two crura?*

Branches of the vagus nerve (oesophageal plexus) pass through the diaphragm with the oesophagus (anterior and posterior gastric nerves).

The muscle fibres of the diaphragm all arch upward to insert into a **central tendon**. This is blended above with the fibrous pericardium.

Qu. 3E *Which large vessel passes through the central tendon of the diaphragm to the right of the midline?*

The diaphragm is supplied by the right and left **phrenic nerves** (which also supply the pleura and peritoneum which cover it). Small branches from the lower six intercostal nerves give sensory fibres to the peripheral region at its attachment to the costal margin. It receives blood from the abdominal aorta via small phrenic arteries; its veins drain.

Structures passing between the thorax and abdomen via the diaphragm

Because the diaphragm completely separates the thoracic and abdominal cavities, the oesophagus, vessels, and nerves which pass between the thorax and abdomen must either pierce the diaphragm or pass between the diaphragm and the body wall.
- The **oesophagus** (p. 85) pierces the diaphragm just to the left of the midline in front of the tenth thoracic vertebra. Although it pierces the left crus, a sling of fibres from the right crus passes around the oesophagus.

Hiatus hernia

Routine contrast radiology of the upper gastrointestinal tract sometimes reveals a small gastric hernia sliding upward through the oesophageal opening of the diaphragm. These 'hiatus hernias' do not usually cause symptoms although 'heartburn' (a burning sensation in the lower central part of the chest) due to reflux of gastric acid into the lower end of the oesophagus, is not uncommon and is often worse when lying down. Large hiatus hernias may enter the pleural cavity and interfere with the normal function of the heart and lungs.

Quadratus lumborum (**6.3.1**) forms the lateral part of the posterior abdominal wall. It arises from the iliolumbar ligament and the posterior part of the iliac crest and passes upward and medially to insert into the tips of the transverse processes of the upper four lumbar vertebrae and into the lower border of the twelfth rib. It is supplied segmentally by the lumbar nerves.

Qu. 3C *What are the actions of quadratus lumborum?*

The **thoraco-lumbar fascia** (**6.3.2**) invests quadratus lumborum and is attached to the transverse processes of the lumbar vertebrae. A more posterior layer surrounds the extensor muscles of the spine and is attached to the spines of the lumbar vertebrae. Laterally, the fascia gives attachment to transversus abdominis and internal oblique.

The left and right **vagus nerves** (which formed the oesophageal plexus) regroup to form the anterior and posterior **gastric nerves** (p. 139) which pass through the hiatus with the oesophagus.

- The **inferior vena cava** pierces the central tendon of the diaphragm just to the right of the midline at the level of the eighth thoracic vertebra. It therefore enters the pericardial cavity to end in the right atrium.
- The **aorta** passes behind the diaphragm and emerges into the abdominal cavity between its two crura in front of the body of the twelfth thoracic vertebra.
- The **thoracic duct** and its dilatation, the **cisterna chyli**, pass behind the diaphragm with the aorta, lying a little to its right.
- The **phrenic nerves** pierce the diaphragm at the margins of the attachment of the pericardium, and supply the diaphragm and its pleural and peritoneal coverings from below.
- The lower six **intercostal nerves** pass into the neurovascular plane of the anterior abdominal wall behind their respective costal cartilages.
- The **sympathetic chain** enters the abdomen between the diaphragm and psoas, under the medial arcuate ligament; the **thoracic splanchnic** (sympathetic preganglionic) **nerves** pierce the crura en route to the coeliac ganglion (p. 127) and the adrenal medulla.

Walls of the pelvis

The walls of the pelvis are formed by the pelvic bone and its associated ligaments and muscles. These give attachment to a basin-shaped muscular septum—levator ani—which forms an actively contractile floor to the pelvic cavity. This is pierced by, and helps control, the outflow tracts of the alimentary, reproductive, and urinary systems.

Because of the orientation of the bony pelvis (p. 105; **6.1.5**) the sacrum forms both its roof and posterior wall, the ischium and part of the ilium form its lateral wall, and the pubis and pubic rami form its anterior wall.

The **sacrotuberous ligament**, attached between the sacrum and ischial tuberosity, and the deeper **sacrospinous ligament**, attached between the sacrum and ischial spine, together convert the bony greater and lesser sciatic notch into the **greater** and **lesser sciatic foramina**.

Piriformis (**6.3.4a**) arises from the inner aspect of the sacrum (second to fourth parts) and passes laterally out of the pelvis through the greater sciatic foramen. **Obturator internus** arises from the inner aspect of the bone surrounding the obturator foramen and from the obturator membrane that almost fills the foramen. It leaves the pelvis through the lesser sciatic foramen. Both these muscles insert into the greater trochanter of the femur. They stabilize and laterally rotate the femur and are supplied by L5, S1, and S2 spinal nerves (Vol. 1, p. 143).

The **floor** of the pelvic cavity is formed by the **levator ani** muscle of either side (**6.3.4b** and see **6.9.11**) and the **pelvic fascia**, which covers the muscle and the vessels supplying the pelvic organs. These form, with the sacrum, a basin-shaped cavity tilted forward toward the pelvic inlet. The forward tilt of the pelvis means that some of the weight of the abdominal contents is taken on the upper surface of the pubis. However, levator ani is important for the integrity of the pelvic floor, and contracts reflexly when abdominal pressure is raised, as in coughing.

The muscular fibres of levator ani arise from the posterior surface of the pubis, from a fibrous tissue arch in the fascia covering obturator internus and from the inner aspect of the ischium and its spine. Some fibres reach the coccyx; others form supporting slings around the ano-rectal junction, vagina, or prostate; many are inserted into the fibrous **perineal body** situated between the anal and urogenital apertures in the pelvic floor.

Vessels on the posterior abdominal and pelvic walls

The **abdominal aorta** is the continuation of the thoracic aorta. It emerges into the abdomen in the midline between the two crura of the diaphragm and passes down on the bodies of lumbar vertebrae to the fourth lumbar vertebra, where it divides into two common iliac arteries (**6.3.5, 6.3.6**). It gives off:

- midline branches to the gut and its derivatives:
  - the **coeliac artery** to the foregut (p. 138);
  - the **superior mesenteric artery** to the midgut (p. 151);
  - the **inferior mesenteric artery** to the hindgut (p. 152);
- paired lateral branches to the organs derived from intermediate mesoderm:
  - **suprarenal arteries** (p. 224);
  - **renal arteries** (p. 171);
  - **gonadal arteries**—testicular (p. 184) or ovarian (p. 194);
- paired branches to the body wall:
  - **phrenic arteries** to the diaphragm;
  - **lumbar arteries** which run laterally over the bodies of the lumbar vertebrae to supply the muscles of the posterior abdominal wall, the back, and the spinal cord (cf. the posterior intercostal arteries; p. 117);
- a single midline terminal branch, the **median sacral artery**, which supplies the lower lumbar and sacral vertebrae and gives small branches to the rectum (p. 152).

The **common iliac arteries** pass downward and laterally above the ala of the sacrum and divide into external and internal iliac arteries in front of the sacroiliac joint. Here the arteries cross anterior to the common iliac veins and sympathetic trunks, but are themselves crossed by the fibres of the presacral plexus

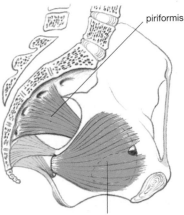

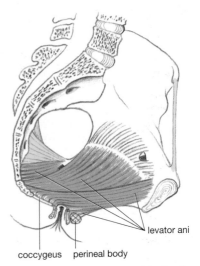

6.3.4 Sagittal section of the pelvis showing (a) muscles and ligaments attached to the inner surface of the pelvic bone; (b) levator ani.

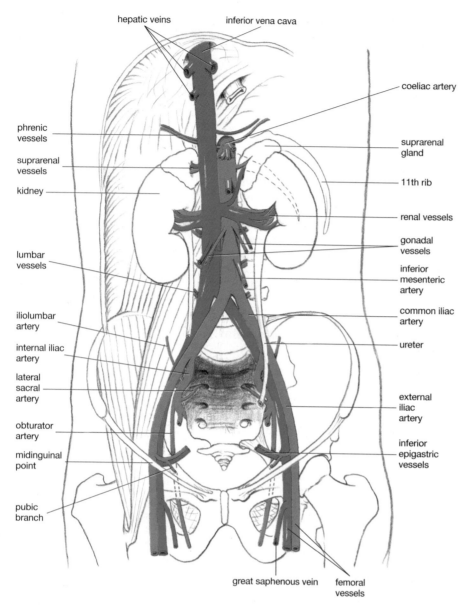

labels on figure:
hepatic veins

inferior vena cava

coeliac artery

suprarenal gland

11th rib

renal vessels

gonadal vessels

inferior mesenteric artery

common iliac artery

ureter

external iliac artery

inferior epigastric vessels

phrenic vessels

suprarenal vessels

kidney

lumbar vessels

iliolumbar artery

internal iliac artery

lateral sacral artery

obturator artery

midinguinal point

pubic branch

great saphenous vein

femoral vessels

6.3.5 Main blood vessels of the posterior abdominal and pelvic walls.

which carries sympathetic fibres down to the pelvic viscera.

Each **external iliac artery** runs along the medial border of psoas to the inguinal ligament. It passes beneath the ligament at the mid-inguinal point (midway between pubic symphysis and anterior superior iliac spine) to enter the thigh as the femoral artery.

Just before the inguinal ligament, it gives off the **inferior epigastric artery**, which ascends medially on the deep surface of the anterior abdominal wall to enter the posterior aspect of the rectus sheath.

Qu. 3F *What is the relation of the inferior epigastric artery to the deep inguinal ring?*

The pubic branch of the inferior epigastric artery (which can replace the obturator artery) may pass over

the mouth of the femoral canal and be involved in a femoral hernia. A cremasteric branch is given off to supply the coverings of the spermatic cord.

Each **internal iliac artery** runs down into the pelvis along the line of the sacroiliac joint. It gives branches to:
• the pelvic viscera;
• the gluteal region;
• the perineum;
• an obturator artery, which runs forward with the obturator nerve over the lateral wall of the pelvis, then leaves the pelvis through the upper part of the obturator foramen to supply the medial aspect of the thigh;
• iliolumbar and lateral sacral arteries, which pass posteriorly to supply skin and muscles of the posterior abdominal wall and lower back.

Veins draining the abdominal and pelvic walls correspond to the arteries (**6.3.5**). The **external iliac** and **common iliac veins** lie medial to and behind their respective arteries; the common iliac veins unite to form the **inferior vena cava** which runs upward on the right side of the lumbar vertebrae with the aorta on its left. Veins draining the gonads, kidneys, and suprarenal glands also drain directly or indirectly into the inferior vena cava.

By contrast, venous blood from the alimentary canal and its derivatives passes almost exclusively to the **hepatic portal vein**, so that the nutrient-rich blood passes through the liver, which controls its content, before being passed via hepatic veins into the inferior vena cava and the systemic circulation.

Veins draining pelvic organs have valveless communications with veins of the lumbar and sacral spine. Malignant deposits from tumours arising in the pelvis can therefore reach the spine by backflow through this low-pressure system.

The major vessels of the abdomen can be outlined by contrast medium. In the arteriogram of the abdominal aorta (**6.3.7**), the cannula through which the contrast medium was inserted is arrowed. 3.22 shows a reconstruction of the vessels from CT. **6.3.8** is a venogram which outlines the major veins.

Qu. 3G *Identify the vessels marked by letters on 6.3.7 and 6.3.8. Why can the branches be followed much further on the arteriogram than on the venogram?*

Lymphatics run with all the major arteries, forming chains of vessels and lymph nodes. Lymph from the lower limbs and pelvis pass, respectively, into the **external iliac** and **internal iliac nodes** (**6.3.9**). These drain upward into a chain of **paravertebral nodes**, which lie laterally on the vertebral bodies and which also drain the suprarenal glands, kidneys and gonads, and posterior abdominal wall muscles.

Lymph from the gut and its derivatives drain along their respective arteries to midline **prevertebral nodes** grouped around the three major arteries (coeliac, superior and inferior mesenteric; see **6.4.11**).

From all these nodes, lymph drains into the **cisterna chyli**, a lymph sac which lies behind the aorta on L1–2 vertebrae and which gives rise to the **thoracic duct** (p. 86).

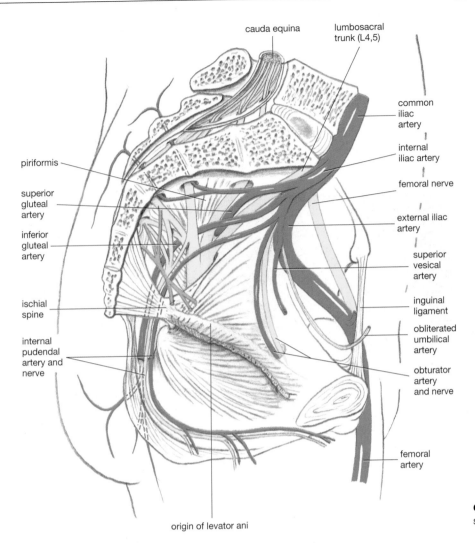

cauda equina

lumbosacral trunk (L4,5)

common iliac artery

internal iliac artery

femoral nerve

external iliac artery

superior vesical artery

inguinal ligament

obliterated umbilical artery

obturator artery and nerve

femoral artery

piriformis

superior gluteal artery

inferior gluteal artery

ischial spine

internal pudendal artery and nerve

origin of levator ani

6.3.6 Arteries and nerves of the pelvis; sagittal view.

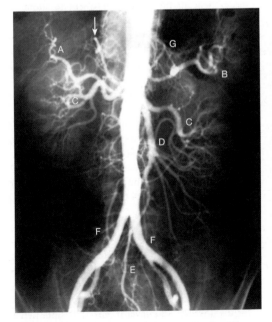

6.3.7 Arteriogram of abdominal arteries (see Qu. 3G; *arrow*, intra-arterial cannula).

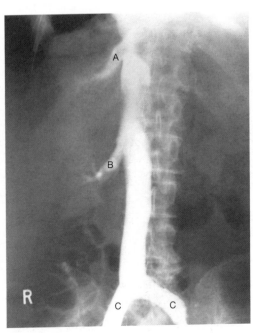

6.3.8 Venogram of abdominal veins (see Qu. 3G).

Lymphangiogram **6.3.9** shows not only external iliac nodes but also inguinal nodes (which lie outside the abdomen and drain the lower limbs).

Nerves on the posterior abdominal and pelvic walls

Somatic nerves

Anterior primary rami of the lumbar spinal nerves form a **lumbar plexus** in psoas major; those from the

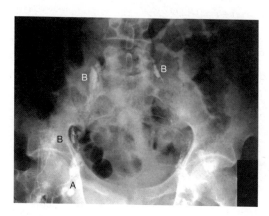

6.3.9 Lymphangiogram of inguinal and abdominal regions. A inguinal nodes; B external iliac nodes.

sacral spinal nerves form a **sacral plexus** on the posterior wall of the pelvis. Together, these are often called the lumbosacral plexus. Most of their branches supply the lower limb and perineum (**6.3.10**).

The **lumbar plexus** has two principal branches:
- The **femoral nerve**, formed from posterior divisions of L2, 3, and 4, runs down the lateral side of psoas to the pelvic brim. It supplies iliacus and then passes beneath the inguinal ligament lateral to the femoral artery to supply the muscles of the anterior compartment of the thigh and a considerable amount of skin in the thigh and leg (Vol. 1, p. 189).
- The **obturator nerve**, formed from anterior divisions of L2, 3, and 4, runs down the medial border of psoas to the obturator foramen, through which it leaves the pelvis with the obturator artery. It supplies parietal peritoneum adjacent to the ovary, so that pain arising in the ovary may be referred to the skin of the medial aspect of the thigh, which the nerve supplies together with the adductor muscles of the thigh (Vol. 1, p. 189).

The lumbar plexus also gives rise to four branches which supply the body wall:
- The **subcostal** (T12), **iliohypogastric** (L1), and **ilio-inguinal** (L1) **nerves** are the abdominal equivalents of the intercostal nerves. They all emerge from the lateral border of psoas, run laterally across quadratus lumborum, then pierce transversus abdominis to pass around the abdominal wall in the neurovascular plane. They supply a strip of skin and muscle around the lower part of the abdomen. The ilio-inguinal nerve supplies muscle strengthening the inguinal canal (p. 116) and the anterior part of the scrotum/labia.
- The **genitofemoral nerve** (L1, 2) is the only nerve to emerge from the anterior surface of psoas. It runs down on its surface and divides. Its femoral branch passes beneath the inguinal ligament to supply a small area of skin over the femoral triangle; its genital branch enters the deep inguinal ring to supply cremaster.

The lumbar plexus is joined to the sacral plexus by the **lumbosacral trunk** (L4, 5), which emerges medial to psoas and runs down over the pelvic brim to join the S1 root of the plexus.

The **sacral plexus** has a number of large branches (superior and inferior gluteal nerves, sciatic nerve, nerves to quadratus lumborum and obturator internus, and the posterior cutaneous nerve of the thigh) which immediately leave the pelvis through the greater and lesser sciatic foramina to supply the lower limb (Vol. 1, p. 188).
- The **pudendal nerve** (S2, 3, 4) leaves the pelvis through the greater sciatic foramen below piriformis, and passes to supply the skin, anal canal, and reproductive and urinary tracts in the perineum (p. 210).
- Within the pelvis, the sacral plexus supplies piriformis (S1, 2) and levator ani (S4) and **pelvic parasympathetic** (S2, 3, 4; pelvic splanchnic) **nerves** to the pelvic viscera (see below).

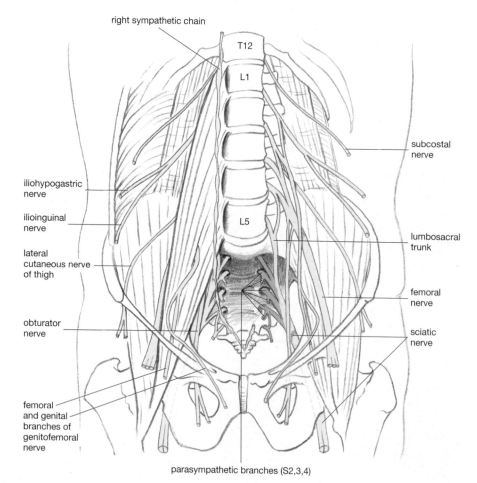

right sympathetic chain

T12

L1

L5

iliohypogastric nerve

ilioinguinal nerve

lateral cutaneous nerve of thigh

obturator nerve

femoral and genital branches of genitofemoral nerve

subcostal nerve

lumbosacral trunk

femoral nerve

sciatic nerve

parasympathetic branches (S2,3,4)

6.3.10 Nerves of posterior abdominal and pelvic walls.

Autonomic nerves (6.3.10, see 2.12, 2.13, 5.5.5)

Sympathetic innervation is vasoconstrictor to the gut and limbs, reduces peristalsis and closes gut sphincters. Branches to the kidney control the vasculature and the secretion of renin; preganglionic branches to the adrenal medulla stimulate the secretion of adrenal catecholamines. In the pelvis, postganglionic fibres inhibit bladder contraction and stimulate contraction of the seminal vesicles and vas deferens (p. 185).

Sympathetic nerves to the abdominal and pelvic viscera are provided mainly by the greater and lesser splanchnic nerves, which are composed of preganglionic fibres branching from the thoracic sympathetic chain (p. 87) and which enter the abdomen through the crura of the diaphragm. They synapse in **ganglia** (coeliac, superior mesenteric, inferior mesenteric, renal) associated with the main branches of the aorta; the postganglionic fibres are distributed to the viscera along the arteries.

The lumbar and sacral parts of the **sympathetic chain** lie anterolateral to the lumbar and sacral vertebrae. They supply the lower part of the alimentary canal (distal to the splenic flexure of the colon) and the pelvic viscera via **presacral nerves** which join the pelvic plexus, and also the posterior abdominal and pelvic walls and the lower limbs.

Parasympathetic innervation stimulates gut motility and secretion, and relaxes sphincters. It stimulates micturition, defaecation, and genital erection. The preganglionic fibres synapse with ganglion cells in the walls of the viscera.

The main parasympathetic supply to the abdominal viscera comes from the **vagus nerves**, which enter the abdomen with the oesophagus. The fibres are distributed largely with branches of the coeliac and superior mesenteric ganglia to the fore- and mid-gut derivatives of the alimentary tract (p. 18).

The **pelvic parasympathetic nerves** (S2, 3, 4; pelvic splanchnic nerves), which arise from the sacral plexus, supply the pelvic viscera and bowel derived from the hind gut. These nerves form a **pelvic plexus** in the fascia around the rectum and pass forward in the pelvic floor to reach the pelvic viscera.

Imaging

Plain radiographs (see **6.1.10, 6.1.11**) show the bony framework of the posterior abdominal and pelvic walls and the margins of psoas. The upper surface of the diaphragm can always be identified because of the contrast provided by the air-filled lungs above; its lower surface can only be seen on a radiograph if gas has escaped from a perforation in the bowel and collected on its under surface, allowing it to be differentiated from the liver below (see **5.2.4**).

The CT **6.3.11** shows an axial section through lumbar vertebra L1. The lowest part of the right crus of the diaphragm, psoas, quadratus lumborum and parts of the 12th, 11th and 10th ribs can be seen. Anterior to the vertebra are the aorta and inferior vena cava.

The axial MRI through a male pelvis (**6.3.12**) shows the pelvic surface of the acetabulum covered with obturator internus, and in the midline, the bladder, the rectum and, between them, the two seminal vesicles and vas deferens. The posterior part of levator ani can be seen as a thin sheet of muscle passing forward from the coccyx to the side wall of the pelvis, separating the pelvis from the ischiorectal fossa.

Coronal MRI (**6.3.13**), through a male pelvis, shows the rectum, bladder, and prostate in the midline and the sloping side wall of the pelvic cavity formed by levator ani covering obturator internus. Within the pelvis, branches of the internal iliac artery can be seen passing to the pelvic organs.

For MRIs of the female pelvis see **6.9.13**.

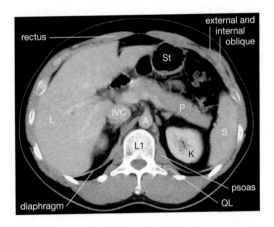

6.3.11 Axial CT through L1, stomach (St), pancreas (P), spleen (S), kidney (K), aorta (A), inferior vena cava (IVC), liver (L), quadratus lumborum (QL).

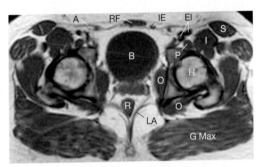

6.3.12 Axial MRI through a male pelvis. Anterior abdominal wall (A), bladder (B), external iliac vessels (EI), gluteus maximus (G Max), head of femur (H), iliacus (I), inferior epigastric vessels (IE), levator ani (LA), obturator internus (O), psoas tendon (P), sartorius (S).

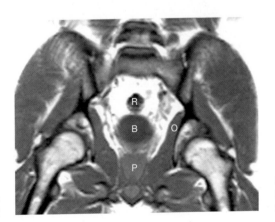

6.3.13 Coronal MRI through a male pelvis. Bladder (B), prostate (P), obturator internus (O), rectum (R).

Questions and answers

Qu. 3A *If pus accumulated within the psoas sheath (e.g. from a tuberculous vertebra), where on the skin would it be likely to discharge ('point')?*

Answer Pus would make its way downward within the psoas sheath to where the muscle is attached to the lesser trochanter of the femur. It would be most likely to point on the skin of the uppermost medial aspect of the thigh.

Qu. 3B *Which action of psoas is shared by iliacus and which is not?*

Answer Hip flexion is shared with iliacus; flexion of the lumbar spine is not.

Qu. 3C *What are the actions of quadratus lumborum?*

Answer Quadratus lumborum produces lateral flexion of the lumbar spine and stabilizes or depresses the twelfth rib, thereby aiding respiration.

Qu. 3D *Which structures enter/leave the abdomen between the two crura?*

Answer Abdominal aorta; thoracic duct.

Qu. 3E *Which large vessel passes through the central tendon of the diaphragm to the right of the midline?*

Answer Inferior vena cava.

Qu. 3F *What is the relation of the inferior epigastric artery to the deep inguinal ring?*

Answer The inferior epigastric artery passes upward, forming the medial boundary of the deep inguinal ring.

Qu. 3G *Identify the vessels marked by letters on 6.3.7 and 6.3.8. Why can the branches be followed much further on the arteriogram than on the venogram?*

Answer On arteriogram **6.3.8**: A-hepatic artery; B-splenic artery; C-right and left renal arteries; D-superior mesenteric artery; E-superior rectal branch of inferior mesenteric artery; F-common iliac artery. On venogram **6.3.9**: A-hepatic vein; B-right renal vein; C-common iliac vein. Arteries can be followed further than veins because contrast material injected into the aorta is distributed by the arterial tree, whereas contrast injected into the venous system will only diffuse back into tributary veins to a very limited extent.

Gastrointestinal system I: General organization and peritoneal attachments; oesophagus, stomach, and duodenum

Gastrointestinal system I: General organization and peritoneal attachments; oesophagus, stomach, and duodenum

This section considers first the development and overall organization of the gastrointestinal tract and its peritoneal attachments within the abdominal and pelvic cavities. The gastrointestinal tract comprises, sequentially: the distal oesophagus; the stomach; the duodenum, jejunum, and ileum, which together form the small intestine; the caecum and appendix, which, with the ascending, transverse, descending, and sigmoid parts of the colon, form the large intestine; the rectum; and the anal canal.

The peritoneum, like the pleura, forms a closed sac that lines the walls of the cavity (parietal peritoneum) and covers much of the tract and its associated organs (visceral peritoneum). It also attaches the tract, directly or indirectly, to the posterior abdominal wall, provides a route whereby vessels and nerves reach the tract, and reduces friction between the moving viscera.

The upper part of the gastrointestinal tract receives ingested food, acts as a sterilizing, macerating reservoir (stomach), and starts digestion and absorption. To aid digestion and absorption, ducts convey into the duodenum bile from the liver and secretions of the pancreas. The bile also conveys certain waste products into the duodenum for disposal in the faeces. The blood draining the tract contains the products of digestion and passes via the hepatic portal vein to the liver. Fat, absorbed as chylomicrons, passes via the lacteals and lymph ducts into the systemic circulation.

Development of the gastrointestinal tract

The epithelium, which lines the alimentary tract from the oropharyngeal membrane to the cloacal membrane, is formed from endoderm. The muscle, connective tissue, and visceral peritoneum, which surround the lining epithelium, are all formed from splanchnic mesoderm. This was formed when the intra-embryonic lateral plate mesoderm divided to produce an inner, splanchnic layer, which surrounds the endoderm, and an outer somatic layer, which lines the overlying ectoderm and forms the parietal peritoneum, muscle, and connective tissue of the abdominal wall.

The cleft between the two layers of mesoderm—the intra-embryonic coelom—becomes the peritoneal cavity. In the abdomen this largely surrounds the gastrointestinal tract (Chapter 4). The cleft does not extend around the dorsal aspect of the endodermal gut tube which, therefore, remains suspended from the dorsal midline of the embryo by a sheet of mesoderm. This forms the dorsal mesentery of the gut and transmits to the gut the vessels derived from the midline dorsal aorta (**6.4.1**) and their associated nerves.

Three distinctive parts of the gut tube develop: the foregut, midgut, and hindgut. In the head and thorax, the cranial part of the foregut forms the pharynx and its derivatives and the oesophagus. Within the abdomen:

- The **foregut** forms the distal oesophagus, the stomach, and the duodenum as far as the entry of the common bile duct, and also the liver and pancreas. Within the abdomen, these are all supplied by the **coeliac artery**, which reaches them through the dorsal mesentery. The foregut is also attached to the anterior abdominal wall by a **ventral mesentery** derived from the septum transversum, in which the liver, biliary system, and ventral pancreatic bud develop.
- The **midgut** forms the caudal part of the duodenum, the jejunum, ileum, and the caecum, appendix, ascending colon, and most of the transverse colon; these are all supplied by the **superior mesenteric artery**.
- The **hindgut** forms the distal third of the transverse colon, descending colon, sigmoid colon, rectum, and upper two-thirds of the anal canal; these are supplied by the **inferior mesenteric artery** (see p. 152).

The lower third of the anal canal develops from the ectodermal **anal pit** and surrounding mesoderm; it is supplied by vessels of the perineum (p. 207).

Although the gastrointestinal tract starts as a midline tube, parts of it undergo rotation and changes of position, so that, in the adult, the tract is anything but in the midline. In addition, some parts of the tract become attached to the posterior abdominal wall in

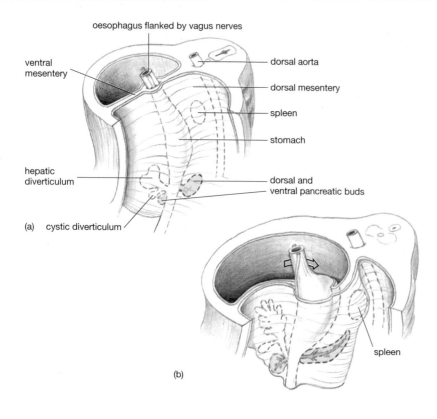

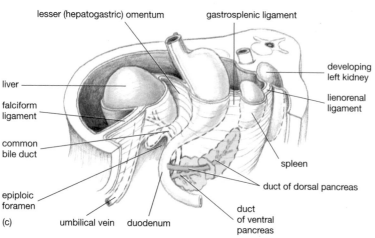

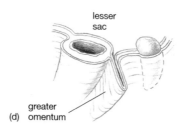

6.4.1 (a), (b), (c) Development of foregut and its derivatives, the liver and biliary system and the pancreas; (d) development of the greater omentum.

their new positions, though such attachments are often incomplete.

The brain, however, continues to interpret **visceral sensation** as if the gut were still a midline tube. All visceral sensation is carried by the autonomic nervous system. Pain from parts of the tract derived from the foregut is 'referred to' (felt in) the epigastrium; pain from the midgut in the umbilical region; and pain from the hindgut in the suprapubic region. Such visceral pain cannot be precisely localized, and patients often indicate its location by moving an entire hand over the area. By contrast, pain arising from the abdominal wall can be located with finger-tip precision because of its somatic nerve supply.

During development, the lumen of both the oesophagus and duodenum become solid cords of endoderm which later recanalize. For this reason congenital atresias and strictures (too little recanalization), and congenital diverticula (too much recanalization) are more common in these than in other parts of the tract. Duodenal atresia is often associated with Down syndrome (a genetic abnormality in which three copies of a part of chromosome 21 are present).

The **stomach** develops from the foregut as a fusiform dilatation, in particular of its dorsal wall. Differential growth carries this dorsal wall and the attached dorsal mesentery toward the left side of the posterior abdominal wall, and the longitudinal axis of the gut rotates clockwise so that the stomach comes to lie somewhat transversely (**6.4.1b, c**). By contrast, the loop of the **duodenum** rotates to the right and, attaching to the posterior abdominal wall, becomes largely retroperitoneal.

From the most caudal part of the foregut, which will form the convexity of the duodenal loop, a hepatopancreatic diverticulum, which will form the liver and ventral pancreas, grows out into the ventral mesentery derived from the septum transversum. As the duodenal loop rotates to the right, this diverticulum migrates posteriorly around to the concavity of the duodenal loop, taking the caudal end of the ventral mesentery with it. Therefore, the common bile duct and pancreatic duct open into the concavity of the duodenal loop. The more proximal part of the ventral mesentery remains as a thin sheet (ventral mesogastrium, **lesser omentum**) between the stomach and the liver.

The dorsal mesentery balloons out to the left and below the developing stomach (**6.4.1d**); the spleen develops in its lateral wall to the left of the stomach. Below the stomach the anterior and posterior walls of the 'balloon' mostly fuse together to form the **greater omentum**, which then becomes attached to the anterior aspect of the transverse colon and its mesentery. Development of the midgut and hindgut is considered with their derivatives on p. 145.

Peritoneum and peritoneal cavity

Within the abdominal and pelvic cavities, the gastrointestinal tract is enclosed within the peritoneal cavity (**6.4.2**).

The **peritoneum** is a thin but firm membrane, similar to the pleura and serous pericardium. It lines the walls of the abdominal and pelvic cavities and covers many viscera. Its moist, smooth surface permits free movement between the viscera and abdominal walls.

As with the pleura, the **parietal peritoneum** lines the walls of the abdominal cavity and is supplied by somatic (intercostal) nerves. Many viscera are covered partly or almost entirely by **visceral peritoneum**, which is supplied only by autonomic nerves.

Qu. 4A *How would pain resulting from irritation of sensory nerves in (a) parietal peritoneum, (b) visceral peritoneum differ?*

Some organs, such as the kidney, lie in contact with the abdominal wall behind the parietal peritoneum; they are said to be **retroperitoneal**. Others, such as the ileum and jejunum, are suspended from the posterior abdominal wall by a double layer of peritoneum—a **mesentery**—in which run vessels and nerves. Yet others, such as the liver, are only partially covered by peritoneum; the areas not covered are called **bare areas**.

The **peritoneal cavity**, like the pleural cavity, is only a potential space containing a thin film of **peritoneal fluid**. It can be subdivided into

- a main part, the **greater sac**, which encompasses most abdominal organs. It is in continuity with the **pelvic cavity** at the pelvic brim;
- a smaller part, the **lesser sac**, which lies largely behind the stomach, and is formed during its embryonic rotation. The opening between the greater and lesser sac is called the **epiploic foramen**.

Situated between the lower surface of the liver and the right kidney is a recess, the **hepato-renal pouch**. The pouch is continuous with a shallow gutter (**right paracolic gutter**) between the ascending colon and the abdominal wall. The corresponding gutter on the left (**left paracolic gutter**) is separated from the lesser sac by the spleen and its peritoneal attachments. Fluid (e.g. from a perforated colon) can collect in these gutters and track down to the pelvic cavity.

An exploring hand can pass freely around the peritoneal cavity at operation, and fluid or gas (e.g. from a perforated viscus) can also spread throughout the cavity. Their distribution is affected by the position of the trunk: fluid will collect at the lowest point, whereas gas will rise. It is important, therefore, to appreciate the

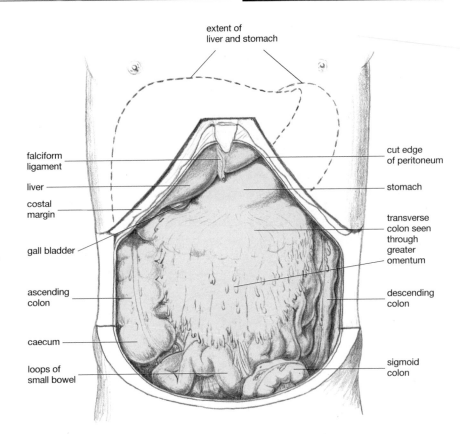

6.4.2 Anterior abdominal wall reflected to show undisturbed viscera.

extent, connections, and barriers between different parts of the peritoneal cavity and the relationship of the peritoneum to different organs.

Qu. 4B *Study radiograph 5.2.4. How was the patient positioned to take this diagnostic radiograph?*

The arrangement of the peritoneum and its reflections from the abdominal walls are shown in **6.4.3**, **6.4.4**, and **6.4.5**.

General arrangement of the gastrointestinal tract and associated viscera

To determine general arrangements of the viscera and their relationships to the peritoneal cavity, the undisturbed contents of an abdomen from which the anterior wall (skin, muscle, fascia, and parietal peritoneum) has been removed (**6.4.2**) should be explored.

In the upper part of an opened abdomen the lower margin of the **liver** can usually be seen with the severed proximal end of the **falciform ligament**—a thin, double sheet of peritoneum joining the liver to the umbilicus and anterior abdominal wall above. Nearly all of the liver is in the right hypochondrium and epigastrium, protected by the ribs; beneath it, the tip of the **gall bladder** appears at the end of the ninth costal cartilage.

The **stomach** occupies most of the epigastrium and left hypochondrium, but varies considerably

---

Peritoneal dialysis

The visceral peritoneum consists of a single layer of squamous epithelium, deep to which are capillaries in a fine interstitial tissue; peritoneum lining the diaphragm contains numerous small lymphatic vessels. This arrangement provides a good potential 'exchange membrane' with an area of about 1 m² in an adult. This potential is used in the technique 'continuous ambulatory peritoneal dialysis' for patients with renal failure. The peritoneal cavity is gently irrigated with a sterile physiological solution. Waste products, particularly urea, cross the peritoneal membrane and are removed with the fluid. This avoids the need to pass the patient's blood through a mechanical dialysis device, as is the case in conventional renal dialysis.

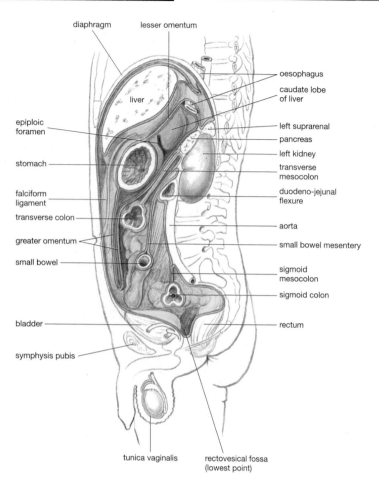

diaphragm

lesser omentum

oesophagus

caudate lobe of liver

liver

epiploic foramen

left suprarenal

pancreas

left kidney

transverse mesocolon

stomach

duodeno-jejunal flexure

falciform ligament

transverse colon

aorta

greater omentum

small bowel mesentery

small bowel

sigmoid mesocolon

sigmoid colon

bladder

rectum

symphysis pubis

tunica vaginalis

rectovesical fossa (lowest point)

6.4.3 Arrangement of peritoneum of the greater and lesser sacs in a parasagittal section of the abdominal cavity.

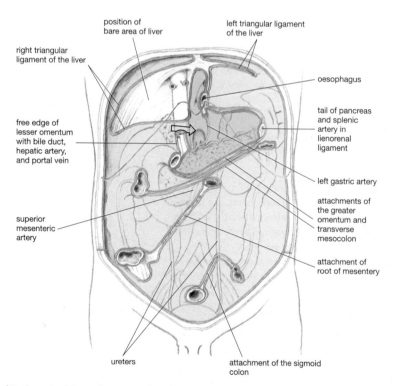

position of bare area of liver

left triangular ligament of the liver

right triangular ligament of the liver

oesophagus

free edge of lesser omentum with bile duct, hepatic artery, and portal vein

tail of pancreas and splenic artery in lienorenal ligament

left gastric artery

attachments of the greater omentum and transverse mesocolon

superior mesenteric artery

attachment of root of mesentery

ureters

attachment of the sigmoid colon

6.4.4 Attachments of the peritoneum to the posterior abdominal wall.

according to its distension and shape. Hidden behind the stomach in the right hypochondrium is the **spleen**.

Hanging down from the lower border of the stomach like an apron is a fatty fold of peritoneum, the **greater omentum**. This covers the transverse colon, which runs horizontally across the abdomen from the liver to the spleen. The greater omentum is variable in extent. If it is short, or lifted up, loops of small intestine (small bowel) will be seen caudal to the transverse colon. The greater omentum is often attached to other parts of the bowel. This is because it acts protectively, wrapping itself around and becoming attached to inflamed areas. Such attachments may, however, provide foci around which the small bowel becomes abnormally twisted (volvulus) and so deprived of its blood supply.

The **duodenum** is relatively inaccessible in the undisturbed abdomen. Its proximal part (the **duodenal cap**) passes to the right from the stomach. It then becomes retroperitoneal, forming a C-shaped loop which is crossed anteriorly by the transverse colon, transverse mesocolon, and the root of the mesentery, which have all been carried to this position by the rotation of the midgut loop. Just to the left of the midline and beneath the greater omentum and transverse mesocolon, the duodenum becomes continuous with the jejunum at the **duodeno-jejunal flexure**.

The loops of small intestine (usually about 6 m in length) fill a large part of the abdominal cavity. They form the **jejunum** proximally and **ileum** distally. Both are attached to the posterior abdominal wall by the fan-shaped **mesentery of the small intestine** (often simply called 'the mesentery').

In the right iliac fossa, the ileum joins a bulbous dilatation of the large bowel, the **caecum**, which is largely covered by peritoneum. On the medial wall of the caecum is attached the blind-ended **vermiform appendix**, which has its own tiny mesentery.

The caecum continues upward in the right side of the abdomen as the **ascending colon**; this is usually more or less attached to the posterior abdominal wall (retroperitoneal). As it reaches the liver, the colon bends sharply (the **hepatic (right colic) flexure**) to become the transverse colon. The **transverse colon** is suspended from the posterior abdominal wall by its own mesentery, the **transverse mesocolon**, which, like the transverse colon, is attached to the greater omentum to form what appears to be a single sheet of omentum (see 6.4.4). In the left hypochondrium, adjacent to the spleen, the colon again makes a sharp bend (the **splenic (left colic) flexure**) to form the **descending colon** which, like the ascending colon, is largely retroperitoneal on the posterior abdominal wall. In the left iliac fossa, the colon acquires another small mesentery (**sigmoid mesocolon**), which demarcates the **sigmoid colon**. The sigmoid colon continues down over the pelvic brim then passes backward beneath the overhanging promontory of the sacrum to its third part, where it becomes continuous with the **rectum**. Here the sigmoid mesocolon ends and the rectum becomes progressively more

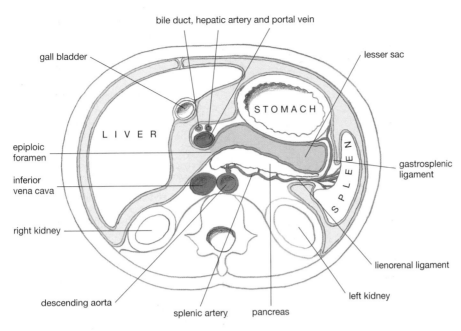

bile duct, hepatic artery and portal vein

gall bladder

lesser sac

STOMACH

LIVER

epiploic foramen

SPLEEN

inferior vena cava

gastrosplenic ligament

right kidney

lienorenal ligament

descending aorta

left kidney

splenic artery   pancreas

6.4.5 Arrangement of the peritoneum of the greater and lesser sacs in a horizontal section of the abdomen through the stomach, viewed from below (cf. CT and MR images).

Oesophageal varices

If there is obstruction to the flow of portal vein blood through the liver, usually as a result of fibrotic scarring (cirrhosis) often resulting from alcohol abuse or viral hepatitis, pressure is raised in the portal system (portal hypertension). As a result the anastomotic veins in the oesophagus, which link the left gastric and azygos veins, become very distended. Whereas endoscopic examination of the normal oesophagus reveals a smooth, pink mucosa (**6.4.6a**), oesophageal varices (**6.4.6b**) appear as linear blue bulges beneath the mucosa. Such distended veins can bleed catastrophically with fatal consequences.

retroperitoneal. Finally, the rectum pierces the pelvic floor to become the **anal canal**.

Oesophagus

The abdominal part of the oesophagus is short; it widens as it passes down to its acutely angled junction with the stomach. It enters the abdomen in front of the tenth thoracic vertebra (at the level of the seventh costal cartilage). Although it pierces the diaphragm just to the left of the midline, it is surrounded by a sling of muscle fibres from the right crus (**5.2.2, 6.3.3**). On contraction, these fibres exert pressure on the diaphragm; this can be observed if a subject takes and holds a deep breath while swallowing a barium meal: the bolus of barium is held up at the point where the oesophagus passes through the diaphragm (see **5.5.4**).

Vessels and nerves of the oesophagus

The lower part of the oesophagus is supplied by branches of the left gastric artery which pass upward through the oesophageal hiatus in the diaphragm. Above the diaphragm oesophageal veins drain into the (systemic) azygos system; below the diaphragm they drain via left gastric veins into the hepatic portal vein. The anastomotic veins joining the two systems lie just beneath the mucous membrane of the oesophagus.

Innervation of the oesophagus is derived from the vagus nerves, which form an oesophageal plexus. As the oesophagus passes through the hiatus in the diaphragm, the plexus regroups to form anterior and posterior gastric nerves.

Oesophago-gastric junction

The lower end of the oesophagus merges with the right border of the stomach at the **oesophago-gastric junction**, making an acute angle (**cardiac notch**) with the fundus of the stomach to its left. The junction is the site of the **cardiac sphincter**. However, although the muscle exerts more pressure here, creating a physiological sphincter, there is no thickening of the circular muscle as there is at the pylorus.

Many activities (e.g. coughing, straining, childbirth) require an increase in intra-abdominal pressure. This will tend to cause reflux of stomach contents into the oesophagus.

Qu. 4C *What mechanisms prevent such reflux, and why is it important that reflux of gastric contents is normally prevented?*

Stomach

The **stomach** (**6.4.7**) is the most dilated part of the alimentary tract (the adult stomach can normally hold up to 1500 ml). It acts as a reservoir which sterilizes, macerates, and starts the digestion of ingested material. Its orientation varies from almost horizontal to vertical. Its shape and position depend on many factors, including body build and posture, and the size and extent of digestion of the last meal. It lies below the lower left ribs, generally overlapping the epigastrium

(a)
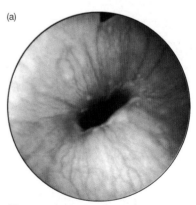

(b)
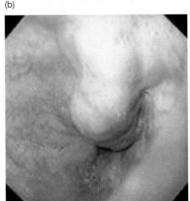

6.4.6 (a) Normal oesophagus; (b) oesophageal varices.

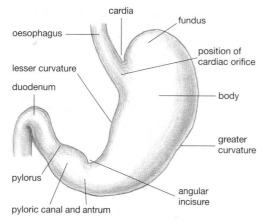

6.4.7 Parts of the stomach.

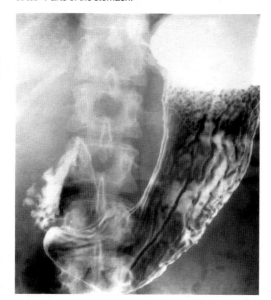

6.4.9 Double-contrast radiograph of stomach showing longitudinal mucosal rugae; the fundus is full of contrast material.

Vomiting

Vomiting is a reflex in which the stomach contents are expelled through the oesophagus and mouth (see **8.4**). Triggered by many different stimuli, it involves powerful contractions of the anterior abdominal wall, diaphragm, and stomach muscle, all of which compress the stomach contents and force them out through a relaxed cardiac sphincter. Other muscles contract to protect the airway so that the vomit is mainly ejected through the mouth.

Congenital pyloric stenosis

Some babies are born with an overactive pyloric sphincter mechanism. This condition is characterized by projectile vomiting after feeding, as the stomach contracts against a closed pylorus. It may require surgical incision of the hypertrophied circular muscle.

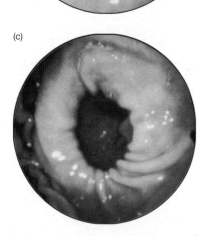

6.4.8 (a):Stomach: rugae in the body of stomach, looking toward the antrum (which lacks rugae) and open pylorus; (b) pylorus closed; (c) pylorus open.

and umbilical regions (see **6.4.17**). The pylorus, which marks the junction of the stomach and duodenum, is rather more constant in position, lying about 1 cm to the right of the midline on the transpyloric plane.

Qu. 4D *Which costal cartilage and which vertebra does the transpyloric plane intersect?*

The right border of the stomach below the entry of the oesophagus forms the **lesser curvature** which may have a sharp **angular notch** before passing upward, backward and laterally to the **pylorus**. To the left of the oesophagus and cardiac notch, the **greater curvature** passes up around the **fundus**—the part of the stomach above the oesophago-gastric junction—then forms the left and lower border of the stomach. The fundus usually contains a gas bubble which can be seen on a plain radiograph of the upright abdomen (see **5.3.11**).

The **body** of the stomach is its largest part. It extends from the fundus to the pyloric region, which begins approximately at a line drawn vertically through the angular notch. The pyloric region consists of the **pyloric antrum** and a short **pyloric canal** (**6.4.8b, c**, see inset **6.4.10**), which is separated from the first part of the duodenum by a ring of fibrous tissue.

Qu. 4E *What are the different functional capabilities of the named regions of the stomach, and how is this reflected in the histology of its mucosa?*

The circular muscle around the pyloric canal is thickened to form the **pyloric sphincter** (see **6.4.8b, c**) which regulates the emptying of the stomach and is controlled by both nerves and hormones. Short, spasmodic contractions propel gastric contents toward the pylorus. If this is closed, the stomach contents become more mixed; if open, a succession of small boluses pass into the duodenum, providing the optimum conditions for digestion and absorption of nutrients.

An erect A-P radiograph will show the presence of gas in the fundus of the stomach (see **5.3.11**), but plain radiographs of the abdomen do not show any other feature of either the normal stomach or duodenum (see **6.1.10**). CT and MRI show the stomach well (see below).

To demonstrate the internal surface radiologically a radiopaque liquid—'barium meal'—must be swallowed. Even better surface detail is revealed if a mixture of barium and an effervescent tablet is swallowed, in order to fill the lumen with both barium and gas—'**double-contrast barium meal**' (**6.4.9, 6.4.10**).

Radiological examination after a barium meal (**6.4.9**) shows that the lining mucosa of the empty stomach is thrown up into folds (rugae; **6.4.8a**), which are most pronounced near the lesser curvature. When drinking fluids, these rugae create a channel which directs the fluid from the oesophagus down the lesser curve and into the pylorus and duodenum. Solid food, however, is directed into the body of the stomach, where it is mixed with and acted on by gastric secretions. Ulcers of the gastric wall

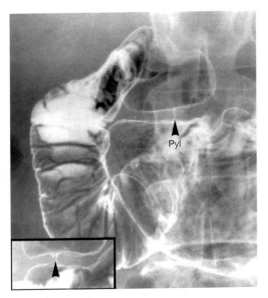

6.4.10 Double-contrast radiograph of pyloric region of stomach and duodenum. The pylorus (Pyl) is closed; in the inset the pylorus (arrow) is a little more open. Horizontal mucosal folds in the duodenum are well defined.

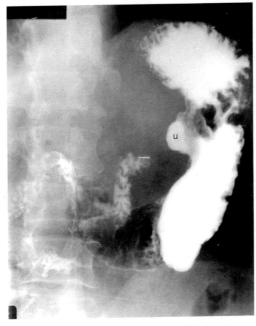

6.4.11 Radiograph of stomach filled with barium, showing a gastric ulcer (U) on the lesser curvature and an abnormal peristaltic wave immediately proximal.

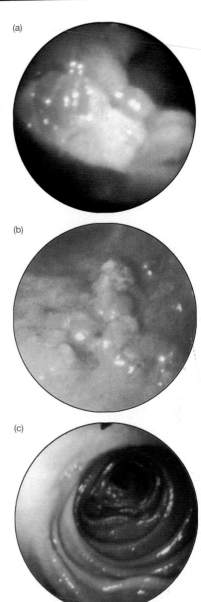

(a)

(b)

(c)

(d)

6.4.12 (a) Gastric ulcer, bleeding; (b) tumour of gastric epithelium; (c) second part of duodenum, viewed from duodenal flexure; (d) duodenal ulcer.

(6.4.11, 6.4.12a) distort the appearance of the stomach wall and mucosa.

Endoscopically, the stomach mucosa should appear smooth and shiny. The rugae (6.4.8a) are often not obvious because the stomach is distended with gas for the procedure. The pylorus can be seen to open and close (6.4.8b, c). 6.4.12a shows a peptic ulcer close to the pylorus; the dead tissue in the ulcer crater has a yellowish appearance. Tumours appear as irregular excrescences of the mucosa (6.4.12b) and both ulcers and tumours may be seen bleeding.

Peritoneal attachments of the stomach

The empty stomach has an anterior and a posterior surface. Like the abdominal oesophagus, these are covered by peritoneum. The anterior surface is covered by peritoneum of the greater sac, the posterior surface by peritoneum lining the lesser sac. At the greater curvature, the peritoneal sheets covering the anterior and posterior surfaces come together to form a two-layered sheet (formed from the dorsal mesogastrium) which is continuous with:

- the **greater omentum**, which is attached to the lower part of the greater curvature;
- a sheet extending laterally from the stomach, ensheathing the spleen, and passing on to the kidney and suprarenal on the posterior abdominal wall. The part between stomach and spleen is called the **gastrosplenic ligament**; that between the spleen and kidney the **lienorenal ligament**. (Peritoneum continuous with the fundus passes back to attach to the diaphragm.)

Because of the rotation of the stomach and its peritoneal coverings, these sheets also form boundaries of the lesser sac (see below).

The **lesser omentum** (formed from the ventral mesogastrium) is attached to the lesser curvature of the stomach and extends up to the under surface of the liver, with a free edge on the right side (6.4.3, 6.4.5) (the original lower border of the ventral mesogastrium). The common bile duct, hepatic artery, and hepatic portal vein all lie between its layers in the free edge (6.6.4). Owing to the rotation of the stomach, they pass vertically between the porta hepatis (the liver hilus) on the under surface of the liver, and the first part of the duodenum, the distal end of which is attached to the posterior abdominal wall.

Behind the free edge of the lesser omentum is the communication (**epiploic foramen**) between the greater sac and the **lesser sac** (6.4.5).

Qu. 4F *The inferior vena cava lies in the posterior wall of the epiploic foramen. What structures limit the foramen above, below, and anteriorly?*

The **lesser sac** extends
- behind the stomach laterally to the spleen and its two peritoneal attachments;
- upward to the liver and behind its caudate lobe to the diaphragm;
- downward behind the gastro-colic part of the greater omentum to the point at which its layers fuse.

An ulcer developing on the posterior wall of the stomach can therefore perforate into the lesser sac, allowing stomach contents to seep into this potential space.

Qu. 4G *If the stomach were removed, which organs would be found in the stomach 'bed' that forms the posterior wall of the lesser sac?*

Duodenum

The **duodenum** (6.4.14) is the most proximal part of the small intestine, in which digestion continues and absorption of most substances starts. It forms a C-shaped loop around the head of the pancreas. Four parts are described:

- The **first** part of the duodenum passes upward, backward and to the right, in front of the common bile duct, portal vein and gastroduodenal artery. On A-P radiographs its oblique course makes it appear foreshortened (the duodenal **cap**). Its proximal part has a small mesentery, but it then becomes retroperitoneal and curves sharply downward.
- The **second, descending** part of the duodenum passes downward retroperitoneally over the medial aspect of the right kidney to the level of the third lumbar vertebra, where it curves to the left. The common bile duct and pancreatic duct enter this part, which is also crossed by the root of the transverse mesocolon (p. 149).
- The **third, horizontal** part of the duodenum crosses the midline of the posterior abdominal wall anterior to psoas, the sympathetic trunk, the inferior vena cava, and the aorta. As it crosses the aorta, it is crossed by the superior mesenteric vessels entering the root of the small bowel mesentery.

- The **fourth** or **ascending** part of the duodenum is short. It passes upward in front of the aorta to the level of the second lumbar vertebra, where it joins the jejunum at the duodeno-jejunal flexure and becomes suspended by the small bowel mesentery. Where the duodenum emerges from its retroperitoneal position, recesses (duodenal fossae) can be present behind the fourth part of the duodenum, and to its left.

6.4.12c shows an endoscopic view of the interior of the second part of the duodenum, viewed from the duodenal cap. The circumferential folds of mucosa (plicae circulares) are characteristic of small bowel mucosa. The common bile duct and pancreatic duct emerge from the head of the pancreas and enter the second part of the duodenum at about its mid-point. (This marks the junction of the embryonic foregut and midgut.) The united ducts form a short **hepatopancreatic ampulla** (ampulla of Vater), and this protrudes through the wall of the duodenum forming the major **duodenal papilla**, which is surrounded by a muscular **sphincter** (of Oddi). About 2 cm proximal to the duodenal papilla the accessory pancreatic duct (if present) enters the duodenum.

Blood supply and lymphatic drainage of the stomach and duodenum

The gastrointestinal tract down to the duodenal papilla is derived from the foregut and therefore is supplied by the **coeliac artery** (6.4.13) which arises from the aorta as it enters the abdomen between the two

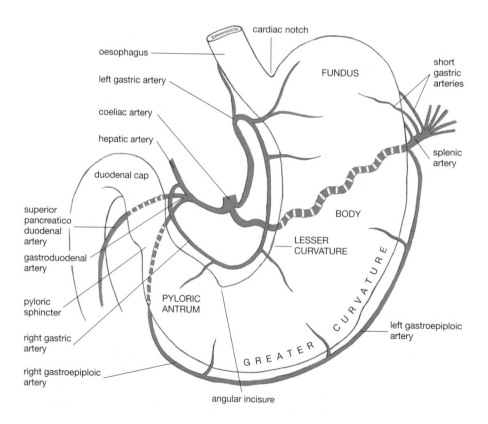

6.4.13 Arterial supply to the stomach.

crura of the diaphragm at the level of the twelfth thoracic vertebral body. The coeliac trunk divides immediately to give rise to:

- The **left gastric artery**, which runs up on the left crus of the diaphragm to the oesophageal hiatus. It gives branches to the oesophagus and then enters the lesser omentum, passing along the lesser curvature to anastomose with the right gastric artery.
- The **hepatic artery**, which passes forward beneath the epiploic foramen to reach the pylorus. Here it gives off a **right gastric artery**, which runs along and supplies the lesser curvature within the lesser omentum, and a **gastroduodenal artery**. The gastroduodenal artery (**6.4.14**) passes down behind the first part of the duodenum, dividing into a **right gastro-epiploic artery**, which runs in the greater omentum to supply the greater curvature of the stomach, and a **superior pancreatico-duodenal artery**, which passes between the duodenum and head of the pancreas, supplying both. (The duodenum below the duodenal papilla is derived from the midgut and therefore supplied by the superior mesenteric artery via an **inferior pancreatico-duodenal artery** (**6.4.14**).)
- The **splenic artery** remains on the posterior abdominal wall, running in an undulating course along the top of the pancreas to enter the lienorenal ligament. Here it divides to supply branches to the spleen, **short gastric arteries** which supply the stomach fundus, and a **left gastro-epiploic artery** which runs along the greater curvature in the greater omentum to anastomose with the right gastro-epiploic artery.

Qu. 4H *In what peritoneal sheet ('ligament') do the short gastric and left gastro-epiploic vessels pass to reach the stomach?*

The stomach is richly vascularized by branches that reach it from the greater and lesser curvatures (**6.4.13**). At their junction is a relatively bloodless plane. Within the stomach wall are many arterio-venous anastomoses.

Qu. 4I *What is the function of these anastomoses?*

The **veins** of the stomach and duodenum correspond to the arteries, except that there is no gastroduodenal or coeliac vein. The vessels drain into the **hepatic portal vein** which runs upward behind the first part of the duodenum into the lesser omentum to ascend to the liver. The right gastro-epiploic vein drains into the superior mesenteric vein, which joins the splenic vein to form the portal vein.

Lymphatics draining the stomach and duodenum run with the arteries to reach nodes around the coeliac artery and nodes at the porta hepatis. Cancer of the stomach can therefore spread to the liver directly through its venous drainage, and also through its lymphatic drainage.

Innervation of the stomach and duodenum

Sympathetic innervation to the stomach (**6.4.15**) and duodenum is derived from the (preganglionic) thoracic

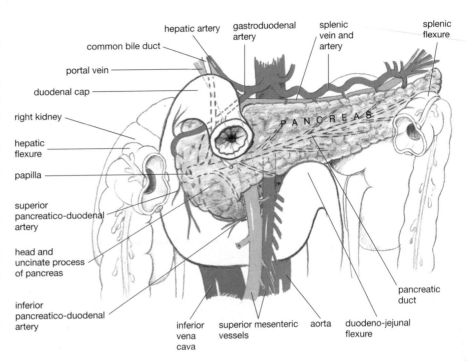

6.4.14 Blood supply to the duodenum and pancreas.

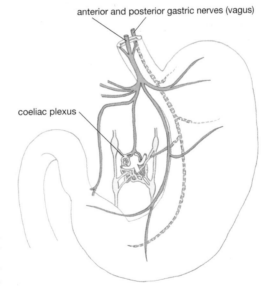

6.4.15 Innervation of the stomach.

splanchnic nerves via the **coeliac plexus**, a mass of fibres and sympathetic ganglion cells lying on either side of the coeliac artery. Postganglionic fibres run along the arterial branches to the viscera.

Qu. 4J *What would be the effects of stimulating the sympathetic nerves to the stomach?*

Parasympathetic innervation to the stomach (**6.4.15**) and duodenum is derived from the **vagus** nerves (which supply all the fore- and midgut derivatives). The oesophageal plexus of the vagus forms anterior and posterior **gastric nerves** which lie anterior and posterior to the oesophagus as it passes through the diaphragm. Both gastric nerves, but primarily the anterior, supply the

Peptic ulceration

Peptic ulceration can occur in either the stomach or the duodenum. It is the result of acid and enzyme attack on the gut wall, caused by overproduction of acid and/or reduced resistance of the mucosa, associated with bacterial (*Helicobacter pylori*) infections, or associated with malignant change of the gastric mucosa. One common site for a peptic ulcer is the first part of the duodenum into which the acidic stomach contents are ejected. **6.4.12d** shows such an ulcer. The area of ulceration is surrounded by a swollen, inflamed area and may be associated with scars from previous, healed ulcers. Peptic ulceration causes pain referred to the epigastric region, and bleeding into the gut, which may be severe if an artery of moderate size is eroded.

Ulcers in any site may perforate the bowel wall and release gas and gut contents into the peritoneal cavity. The gas can collect beneath the diaphragm where it can be detected radiologically (see **5.2.4**). The acid is intensely irritant to the peritoneum and causes reflex rigidity (guarding) of the anterior abdominal wall muscles. A perforation through the posterior wall of the stomach will erode into the lesser sac. If the epiploic foramen becomes occluded, a fibrotic cyst can form in the lesser sac.

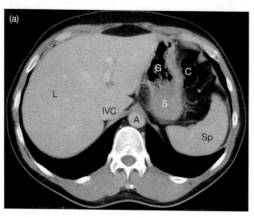

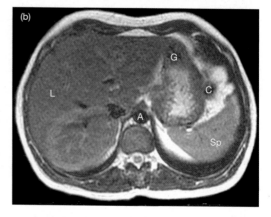

6.4.16 (a) Axial CT and (b) axial MRI of abdomen, showing stomach (S) containing gas (G). Aorta (A), splenic flexure of colon (C), inferior vena cava (IVC), liver (L), spleen (Sp).

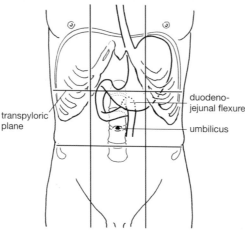

6.4.17 Surface markings of the stomach and duodenum on the anterior abdominal wall. The pylorus lies in the 'trans-pyloric plane'.

stomach. Parasympathetic nervous activity increases gastric acid secretion and, before the discovery of effective drugs (histamine H_2 antagonists), surgical division of the vagal supply to the stomach was used to control excessive acid secretion. However, the operation had undesirable effects if vagal fibres to the pylorus, liver, and coeliac plexus were divided.

Qu. 4K *What effect would cutting the vagus have on gastric movements and tone in the pyloric sphincter?*

Computerized imaging of the stomach and duodenum

CT and MRI sectional images (**6.4.16** and see **6.11.1**) reveal various parts of the stomach and duodenum without the need to fill the organs with barium. As the subject is lying supine, gas in the stomach collects in the area beneath the anterior abdominal wall, forming a fluid level interface with the contents.

Living anatomy

To fix in your mind the surface relations of the internal organs, it is helpful to draw these and the lines marking the regions on your own or a colleague's abdomen (**6.4.17**). The positions of the oesophagus in the oesophageal hiatus of the diaphragm, and of the second, third and fourth parts of duodenum are rather constant; the position and orientation of the stomach varies considerably.

Percussion of the left hypochondrium should locate an area of resonance caused by the gas in the fundus of the stomach. A stethoscope placed over the stomach will reveal sounds created by gastric movements.

Questions and answers

Qu. 4A *How would pain resulting from irritation of sensory nerves in (a) parietal peritoneum, (b) visceral peritoneum differ?*

Answer The parietal peritoneum is innervated by somatic sensory nerves supplying the abdominal wall. This allows localization of pain with finger-tip precision.

The visceral peritoneum is supplied by sensory fibres running with sympathetic (and parasympathetic) nerves to the viscera. The central nervous connections of these sensory fibres allow pain to be recognized but not very precisely localized, so a patient is likely to indicate its position with the flat of the hand.

Qu. 4B *Study radiograph 5.2.4. How was the patient positioned to take this diagnostic radiograph?*

Answer The patient must be sitting or standing erect because the gas has risen to the undersurface of the diaphragm.

Qu. 4C *What mechanisms prevent such reflux, and why is it important that reflux of gastric contents is normally prevented?*

Answer Reflux is prevented partly by the cardiac sphincter, partly by the angle at which the oesophagus enters the stomach, and partly by the sling of fibres from the right crus of the diaphragm, which gives additional support when intra-abdominal pressure is raised. The mucosa of the oesophagus is not resistant to gastric acid. Therefore, if reflux occurs, the mucosa becomes inflamed, causing midline pain behind the sternum—'heartburn'. The inflammation can also cause ulceration and scarring, which narrows the oesophagus, causing difficulty with swallowing.

Qu. 4D *Which costal cartilage and which vertebra does the transpyloric plane intersect?*

Answer The transpyloric plane is a horizontal plane situated midway between the suprasternal notch and pubic symphysis. It transects the tip of the ninth costal cartilage and the lower border of L1 vertebra.

Qu. 4E *What are the different functional capabilities of the named regions of the stomach, and how is this reflected in the histology of its mucosa?*

Answer The mucosa of the fundus produces mainly mucus. The body of the stomach produces mucus, acid (from parietal cells), enzymes, particularly pepsin (from chief cells), and some hormones (from gut endocrine cells); the pyloric antrum produces mucus and a number of hormones, particularly gastrin and somatostatin.

Qu. 4F *The inferior vena cava lies in the posterior wall of the epiploic foramen. What structures limit the foramen above, below, and anteriorly?*

Answer The epiploic foramen is limited above by the portal hepatis; anteriorly by the free margin of the lesser omentum with its contained vessels (common bile duct, hepatic artery, hepatic portal vein); and below by the first part of the duodenum and the hepatic artery as it passes from the posterior abdominal wall into the lesser omentum.

Qu. 4G *If the stomach were removed, which organs would be found in the stomach 'bed' that forms the posterior wall of the lesser sac?*

Answer In the upper part of the stomach bed are the left suprarenal gland, upper pole of the left kidney, and spleen; below lie the body and tail of the pancreas with the transverse mesocolon attached to them; behind the transverse mesocolon lies the duodeno-jejunal flexure and coeliac ganglion.

Qu 4H *In what peritoneal sheet ('ligament') do the short gastric and left gastro-epiploic vessels pass to reach the stomach?*

Answer The splenic artery leaves the posterior abdominal wall enclosed in the peritoneum which forms the lateral wall of the lesser sac—the lienorenal 'ligament'. From the spleen its branches pass to the stomach via the gastrosplenic 'ligament'. Both 'ligaments' are formed medially by peritoneum bounding the lesser sac and laterally by peritoneum bounding the greater sac.

Qu 4I *What is the function of these anastomoses?*

Answer The arterio-venous anastomoses bypass blood from the gastric capillary circulation when splanchnic perfusion is decreased; they thereby help to maintain pressure in the hepatic portal venous system.

Qu. 4J *What would be the effects of stimulating the sympathetic nerves to the stomach?*

Answer Stimulation of sympathetic nerves to the stomach produces vasoconstriction, a reduction in peristalsis, and increased tone in the pyloric sphincter.

Qu. 4K *What effect would cutting the vagus have on gastric movements and tone in the pyloric sphincter?*

Answer Gastric movements would be inhibited and the tone in the pyloric sphincter increased.

Gastrointestinal system II: Small and large intestines

Gastrointestinal system II: Small and large intestines

The distal parts of the small intestine (jejunum and ileum) complete the processes of digestion and absorption of nutrients. Soluble absorbed nutrients are carried by the hepatic portal vein to the liver where they are metabolized, stored, or released into the systemic circulation. Emulsified lipids are carried from the small intestine by lymphatics (lacteals) which drain into the systemic circulation via the thoracic duct.

The large intestine (caecum, ascending, transverse, descending, and sigmoid colon) absorbs water and ions, thereby converting the liquid contents from the small intestine into semisolid faeces. Faecal material is stored predominantly in the sigmoid colon. From there it is moved into the rectum, which acts as a sensor and temporary reservoir before the faeces are extruded through the anal canal. Sphincters around the anal canal maintain faecal continence prior to defaecation.

Development of the intestines

As the stomach and lesser sac are extending to the left and rotating clockwise, the **midgut**, which forms the bowel from the duodenal papilla to the splenic flexure of the colon, lengthens rapidly to form a large loop attached to the posterior abdominal wall by a dorsal mesentery. The superior mesenteric artery lies in the axis of this loop (**6.5.1a**). The apex of the loop is initially continuous with the vitello-intestinal duct, but this normally disappears entirely.

Growth of the gut and other intra-abdominal organs proceeds more rapidly than the abdominal cavity can accommodate, and the midgut loop herniates into the extra-embryonic coelom within the developing umbilical cord. Here its proximal limb and the adjacent part of the distal limb elongate greatly, forming coils of small intestine, while the rest of the distal limb increases in diameter but elongates less markedly, to form the caecum, ascending, and transverse colon. The parts of the gut tube that will form the duodeno-jejunal flexure and the splenic flexure of the colon are prevented from herniation by 'retention bands' attached to the posterior abdominal wall.

Before and during its return to the umbilical cavity, the midgut loop undergoes an anticlockwise rotation of ~270° about an axis formed by the superior mesenteric artery in the dorsal mesentery (**6.5.1b**). The apex of the loop is the vitello-intestinal duct, which should regress completely, but which may remain, usually as a Meckel's diverticulum, attached to the side of the terminal ileum. This rotation carries the distal limb of the loop that will form the caecum, ascending and transverse colon, upward and then to the right of the proximal limb, which will form most of the small bowel.

The proximal limb re-enters the abdominal cavity first, pushing that part of the **hindgut** that will form the descending and sigmoid colon over to the left, where the descending colon and its mesentery become fused to the posterior abdominal wall. The sigmoid colon retains a small mesentery.

The transverse colon retains a large mesentery, the transverse mesocolon, which is slung from the posterior abdominal wall along the lower border of the body of the pancreas. The ascending colon and caecum pass down to the right iliac fossa, and the ascending colon and its mesentery become attached to the posterior abdominal wall in the same way as the descending colon.

The superior mesenteric artery, about which the midgut loop rotates, comes to lie in the line of attachment of the small bowel mesentery to the posterior abdominal wall. This is because the dorsal mesentery of most of the distal limb of the loop has become attached to the posterior abdominal wall, while that of the proximal limb of the loop remains largely free, forming the mesentery of the small intestine.

The expansion of the dorsal mesogastrium that forms the greater omentum grows down over the anterior aspect of the transverse colon and its mesocolon and becomes attached to both.

Incomplete rotation or malrotation of the gut may occur. For example, the caecum and appendix may be placed high up beneath the right lobe of the liver. Rarely, the gut rotates clockwise and the positions of the ascending and descending colon are reversed, the caecum and appendix lying in the left iliac fossa instead of the right.

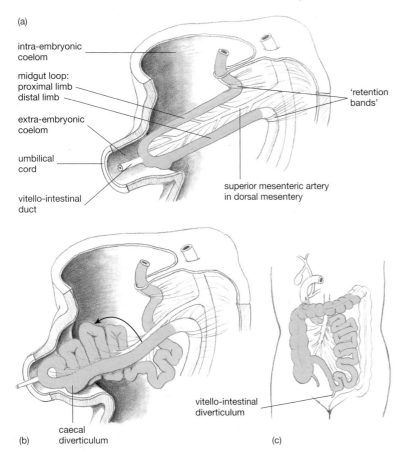

(a)

intra-embryonic coelom

midgut loop: proximal limb distal limb

extra-embryonic coelom

umbilical cord

vitello-intestinal duct

'retention bands'

superior mesenteric artery in dorsal mesentery

caecal diverticulum

(b)

vitello-intestinal diverticulum

(c)

6.5.1 Development of the proximal and distal limbs of the midgut loop.

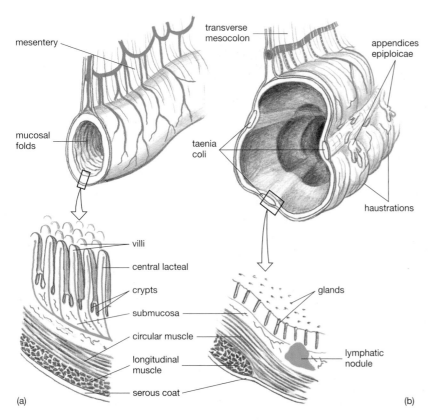

mesentery

mucosal folds

transverse mesocolon

appendices epiploicae

taenia coli

haustrations

villi

central lacteal

crypts

submucosa

circular muscle

longitudinal muscle

serous coat

glands

lymphatic nodule

(a)

(b)

6.5.2 Differences between (a) small bowel and (b) large bowel.

Small intestine: jejunum and ileum

The small intestine (small bowel) is a continuous tube (~280 cm in an adult) comprising the duodenum (p. 134), jejunum, and ileum. Its principal function is the completion of the digestion and absorption of most nutrients. Both the jejunum and ileum are lined with a highly specialized mucosa, thrown into circular folds and numerous villi covered with microvilli, all of which increase the surface available for digestion and absorption.

The **duodeno-jejunal flexure** marks the start of the jejunum and the **ileo-caecal junction** marks the end of the ileum, but there is no obvious point of demarcation between the jejunum and ileum. The jejunum, however, tends to be of greater diameter than the ileum and to have a thicker mucosa, with more circular folds, which can be appreciated by palpation of the living bowel. Also, the ileum, but not the jejunum, has aggregates of lymphoid tissue—Peyer's patches—which are visible and palpable at the antimesenteric border and are concerned with the formation of antibodies against foreign proteins. Other differences between the mucosa of jejunum and ileum require microscopic examination.

The jejunum and ileum are both suspended from the posterior abdominal wall by the fan-shaped **small bowel mesentery** (6.5.2, 6.5.3), which is attached to the posterior abdominal wall (see **6.4.4**) along a line that runs from the duodeno-jejunal flexure downward toward the right iliac fossa, where the caecum lies and the ascending colon begins. It contains all the vessels and nerves passing to and from the small bowel.

Qu. 5A *How could you use the mesentery to determine, at operation, which is the proximal end of a loop of bowel which presented through an incision?*

The lumen of the small intestine cannot be visualized by conventional radiology unless it contains either radiopaque contrast medium or gas. Small amounts of swallowed air do occur in the small bowel, but large amounts are only found there if bacteria have entered

---

Meckel's diverticulum

A **Meckel's diverticulum** is not a common developmental anomaly but, if one is present, it protrudes for a variable length from the antimesenteric border of the ileum about 60 cm from the ileo-caecal junction. The diverticulum may be connected to the umbilicus by a fibrous cord around which a loop of intestine can twist and become strangulated. Such a connection may have a lumen, so that the ileum and umbilicus are connected by a fistula. The mucosa in a Meckel's diverticulum varies—presumably reflecting a lack of differentiation 'instructions' in an area of bowel that does not normally develop. If gastric mucosa is formed, a peptic ulcer may develop in the adjacent ileum.

Qu. 5B *What is a Meckel's diverticulum a remnant of?*

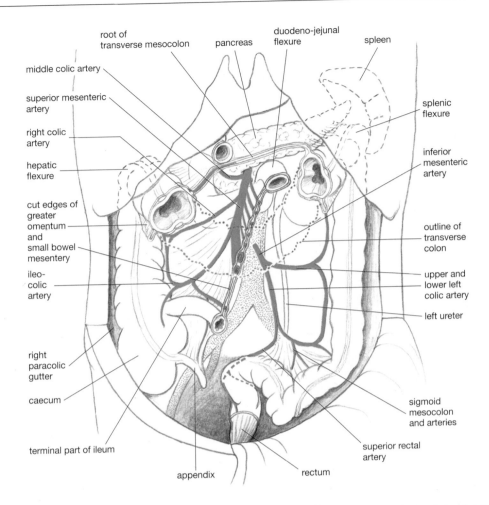

6.5.3 Arterial supply to the small and large bowel derived from the superior and inferior mesenteric arteries.

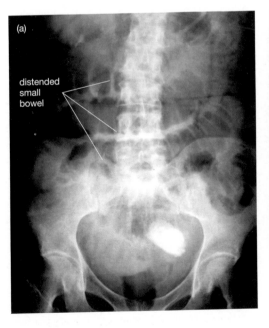
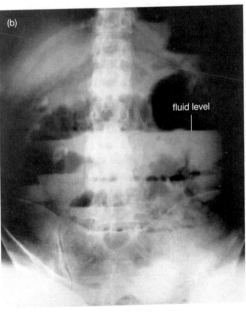

6.5.4 Double-contrast radiographs of small bowel in a patient with intestinal obstruction. The small bowel is distended with gas. In (a) the patient is supine, and contrast medium and gas outline the transverse mucosal folds. In (b) the patient is upright and the gas–liquid interface forms fluid levels.

the small bowel from the large bowel as a result of obstruction of the large bowel. Because small bowel contents are fluid, the presence of a large amount of gas creates numerous fluid levels in the coils of small intestine if the patient is upright (**6.5.4b**) but not when supine (**6.5.4a**). However, the supine radiograph shows more clearly the extent of gas in the small bowel and, hence, the position of the obstruction.

Introduction of barium into the small bowel by a barium meal or via an endoscope (**6.5.5**) reveals the

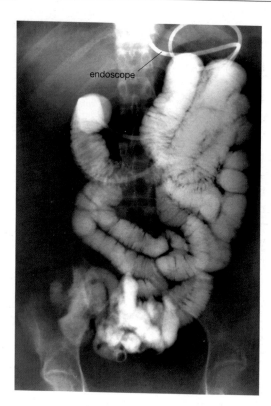

6.5.5 Radiograph showing details of small bowel, in particular the transverse mucosal folds. Contrast medium has been introduced via an endoscope.

interior of the small bowel and the pattern of its mucosa.

Prior to the introduction of fibre-optic instruments, the interior of the small bowel could not be viewed directly. Long fibre-optic endoscopes can now be passed from the mouth to the small bowel, and also retrogradely from the anus. These are used to inspect the bowel and, if necessary, take biopsies. The endoscopic appearance of the normal small intestine is like that of the duodenum (see **6.4.15**).

Ileo-caecal junction

The ileum opens on to the postero-medial wall of the caecum as a horizontal slit, surrounded by two folds of mucosa which protrude (like lips) into the caecum (**6.5.6**). These form a valve-like structure but this does not prevent some caecal contents from entering the terminal ileum, particularly if pressure in the caecum is raised.

Large intestine

The large intestine comprises the caecum and its appendix, the ascending, transverse, descending and sigmoid colon, and the rectum. It extends from the ileo-caecal junction to the anal canal in the pelvic floor. It can be distinguished from small bowel (**6.5.2**) by:
- its larger calibre;
- its longitudinal muscle layer which is divided into three bands—taenia coli—which extend from the caecum to the beginning of the rectum (the rectum and appendix have continuous longitudinal muscle coats);
- sacculations of its wall, known as **haustrations**, which are produced by the relatively shorter length of the taeniae;
- the presence of fat-filled tags—appendices epiploicae—which are attached to the ascending, transverse, descending, and sigmoid colon.

The **caecum** is the blind-ending dilated part of the large bowel extending below the ileo-caecal junction. Its posterior wall may be attached to the right iliac fossa but, more usually, is separated from it by a **retro-caecal fossa** which, if extensive, can extend upward as far as the liver, and which frequently contains the appendix.

The vermiform **appendix** (**6.5.6**) is attached to the medial wall of the caecum and has a complete longitudinal muscle wall formed by the confluence of the three taenia. It is very variable in both length and position, lying usually either
- behind the caecum (retrocaecal), or
- hanging over the pelvic brim (pelvic).

The base of the appendix is rather constant in position; its surface marking (McBurney's point) is the junction of the middle and lower thirds of a line joining the anterior superior iliac spine and the umbilicus. This is an important surgical landmark (see **6.2.8**).

The lumen of the appendix is very narrow; swelling of the large amount of lymphoid tissue in its wall can easily obstruct it.

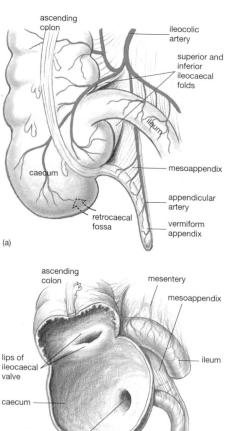

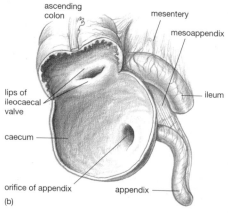

6.5.6 (a) Ileo-caecal junction, caecum, and appendix with its mesentery and arterial supply. (b) Internal aspect of caecum, showing the ileo-caecal valve and appendicular orifice.

Appendicitis and allied problems

Appendicitis—inflammation of the appendix—is a common condition, especially in children. It is usually caused by inflammation of the extensive lymphoid tissue in the appendix wall, which can block the lumen. At first, the inflammation involves the visceral peritoneum and causes pain which is poorly localized in the region of the umbilicus, loss of appetite, and often vomiting. However, when the inflammation spreads to parietal peritoneum the somatic pain produced is much sharper and more localized. If the appendix touches the anterior abdominal wall, the pain is localized to the right iliac fossa. Even before this happens, gentle pressing of the abdominal wall over the root of the appendix can elicit tenderness.

Inflammation of a retrocaecal appendix can spread to the parietal peritoneum overlying psoas and iliacus, causing pain in the right loin and reflex spasm of the muscles with flexion of the right hip.

Inflammation of a pelvic appendix is often misdiagnosed because it can involve the ureter, ovary and ovarian tube, or even the rectum, and thus mimic urinary infection, an ovarian cyst, or tubal infection, and cause diarrhoea.

Other conditions that can be confused with appendicitis include: inflammation of mesenteric lymph nodes in the small bowel mesentery (mesenteric adenitis); any perforation of the bowel that affects the right iliac fossa; inflammation of a Meckel's diverticulum; torsion of the ovary; or an ectopic pregnancy.

The appendix has a number of peritoneal attachments:

- a **mesoappendix** or mesentery, which connects it to the mesentery of the ileum and in which an appendicular artery runs to the tip of the appendix;
- an inferior ileo-caecal fold ('bloodless fold of Treves'), which may fuse with the mesoappendix;
- a superior ileo-caecal fold, in which runs an anterior caecal artery (6.5.6).

These folds produce a number of peritoneal fossae.

The **ascending colon** runs from the right iliac fossa to the hepatic (right colic) flexure beneath the right lobe of the liver. Lateral to the ascending colon, the **right paracolic gutter** (6.5.3) connects the hepato-renal pouch above and pelvic cavity below. The ascending colon is usually closely attached to the posterior abdominal wall (i.e. retroperitoneal) but, like the descending colon, in about 25% of people the attachment to the posterior abdominal wall is incomplete and the ascending colon may even have a partial mesentery.

Qu. 5C *On which muscles and what organs does the ascending colon lie?*

The **hepatic flexure** of the colon lies between the right lobe of the liver, from which it is separated by the peritoneal cavity, and the right kidney, over which it is attached, at the level of the tenth right costal cartilage.

The **transverse colon** with its mesentery forms a loop between the hepatic and the splenic flexure, which hangs down to a variable extent. The **transverse mesocolon** is attached to the posterior abdominal wall over the right kidney, descending duodenum, and the head, body, and tail of the pancreas. It divides the greater sac of the peritoneal cavity into a **supracolic** and an **infracolic** compartment. The middle colic branch of the superior mesenteric artery enters its root and supplies the transverse colon. The greater omentum lies over and is attached to much of the transverse mesocolon and transverse colon.

The **splenic flexure** is usually much more acute and placed a little higher than the hepatic flexure (behind the eighth left costal cartilage, in contact with the left kidney, tail of pancreas, spleen, and greater curvature of the stomach). At this point the colon is attached to the diaphragm by the phrenico-colic ligament, a fold of peritoneum which, in a recumbent patient, can present a barrier between the **left paracolic gutter** below and the supracolic compartment of the peritoneal cavity above.

The **descending colon** passes from the splenic flexure to the left iliac fossa and, in most people, is attached to the posterior abdominal wall (retroperitoneal).

Qu. 5D *From which part of the primitive gut tube is the descending colon derived, and therefore which artery supplies it?*

The **sigmoid colon** forms a loop of variable length extending from the left iliac fossa to the beginning of the rectum, which starts at the third piece of the sacrum in the midline. Faeces are stored in the sigmoid colon until defaecated. It is suspended by the **sigmoid mesocolon** as a loop, or loops, which hang down into the pelvis. The attachment of the sigmoid mesocolon to the posterior abdominal wall forms an inverted V, the apex of which overlies the left sacroiliac joint, at the point where the ureter crosses the left common iliac artery; its left limb follows the pelvic brim, its right limb passes down to the third sacral vertebra.

The **rectum** follows the concavity of the sacrum and coccyx as they form the curved roof and posterior wall of the pelvis. It begins and ends in the midline, but its middle third bulges to the left side. Internally, these curves create three horizontal folds of muscle and mucosa—the **rectal valves**. Its lower part, the **ampulla of the rectum**, is distensible but, like the rest of the rectum, normally contains faeces only immediately before defaecation. As the rectum passes through levator ani it becomes continuous with the anal canal.

The rectum has a complete longitudinal smooth muscle coat. Peritoneum clothes its upper third anteriorly and laterally, but only the anterior surface of its middle third. The lower third of the rectum is not in contact with peritoneum, which is reflected forward from its middle third on to the bladder in males, and on to the posterior fornix of the vagina and uterus in females. In this way either a **recto-vesical pouch** or a **recto-uterine pouch** (pouch of Douglas) is formed. These are the most

Cancer of the colon

The second most common cancer of the Western world is that of the large bowel (**6.5.7**). There are striking differences in the incidence of this disease, which suggest an environmental cause. Seventy per cent of colon cancer involves the descending and sigmoid colon, with only 20% in the ascending colon. The disease presents with a range of symptoms, which depend in part on the anatomical structures in the vicinity of the tumour, but which usually includes the presence of blood in the faeces.

Cancer of the caecum is often 'silent' at first and may cause anaemia due to chronic blood loss. At the ileo-caecal junction bowel contents are liquid and bowel obstruction therefore develops only slowly, but if the tumour occludes the lumen of the appendix it may cause appendicitis. In the descending and sigmoid colon the faeces are firm, and a tumour can therefore obstruct the bowel more easily. If the tumour invades the wall of the bowel, it can form attachments to and invade adjacent viscera, such as the bladder and/or uterus, so that, in extreme cases, faecal material can appear in the urine or vagina.

Qu. 5E *Where might secondary deposits from cancer of the colon be found?*

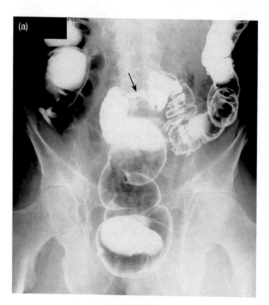

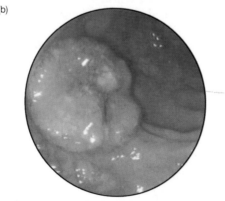

6.5.7 (a) Double-contrast radiograph showing the 'apple-core' appearance (arrow) caused by a tumour of the sigmoid colon. (b) Endoscopic view of cancer of colon.

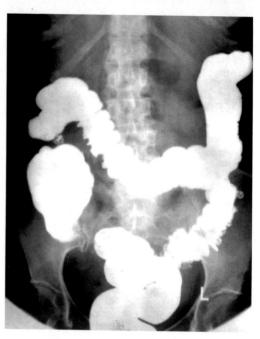

6.5.8 Radiograph taken after a barium enema which fills the colon and caecum. The appendix is outlined by barium because other contents fill its lumen.

dependent parts of the peritoneal cavity and therefore fluid or pus accumulating in the peritoneal cavity will flow into them unless prevented by inflammatory adhesions.

Ano-rectal junction and anal canal

The **ano-rectal junction** lies just in front of the tip of the coccyx and, at this point, the bowel narrows abruptly and curves sharply posteriorly.

Below the ano-rectal junction the **anal canal**, 3–4 cm in length, passes downward and backward to the **anus**.

Its acute angulation at the recto-anal junction is maintained by the forward pull of a muscular sling (pub-orectal sling) derived from levator ani which passes posterior to the bowel and has a sphincter-like effect.

The anal canal is guarded by two sphincters: an inner involuntary sphincter formed by the continuation of the circular smooth muscle of the rectum; and an external, voluntary sphincter of striated muscle. In addition, specializations of the submucosa—the anal cushions—help maintain continence to fluid and flatus.

The anal canal, defaecation, and faecal continence are dealt with in more detail on p. 207.

The large bowel naturally contains gas, which permits it to be visualized to some extent on a plain radiograph (see **3.3**, **6.1.10**). If, however, the large bowel needs to be examined, it must first be cleared of faeces and then filled with radiopaque material (**6.5.8**) usually together with gas (double contrast: **6.5.9**, **6.5.10**) by means of an enema.

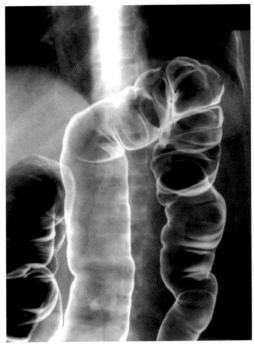

6.5.9 Double-contrast radiograph, showing the angulation of the splenic flexure of colon.

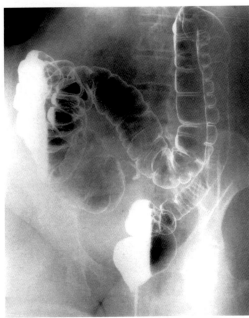

6.5.10 Double-contrast radiograph of large bowel (patient lying on right side).

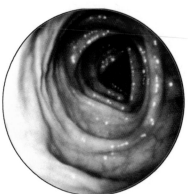

6.5.11 Endoscopic view of transverse colon.

Qu. 5F *How does the appearance of the colon differ from that of the small bowel when filled with contrast medium?*

In the past, the sigmoid colon was the furthest part of the large bowel that could be viewed by use of a long, straight tube (sigmoidoscope), although often only the rectum could be examined because of an acute angulation of the recto-sigmoid junction. Modern flexible endoscopes can view the entirety of the large bowel. Its interior varies: the walls of the rectum are smooth apart from the rectal valves; haustrations are seen in the remainder of the colon, and the transverse colon has a characteristic triangular-section lumen (**6.5.11**).

Qu. 5G *What is the reason for this triangular appearance of the transverse colon?*

Blood supply and lymphatic drainage of the small and large intestines

The **superior mesenteric artery** supplies the midgut-derived bowel from the duodenal papilla to the splenic flexure with a series of anastomosing vessels. It arises from the aorta in front of the body of the first lumbar vertebra and behind the neck of the pancreas, and crosses the third part of the duodenum to enter the root of the small bowel mesentery. In developmental terms, it ends at the apex of the midgut loop, i.e. the ileum proximal to the ileo-caecal junction, where a Meckel's diverticulum may persist. Its anastomosing branches are:

- the **inferior pancreatico-duodenal artery**—a small branch given off almost immediately (p. 139);
- numerous **jejunal** and **ileal** branches, which arise from the left of the superior mesenteric artery, passing into

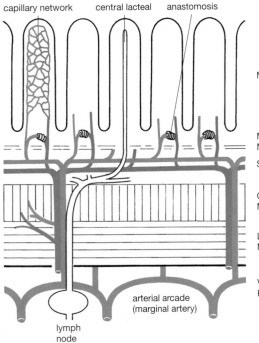

6.5.12 Diagram of blood supply to the wall of the intestine.

the mesentery to form anastomosing **arcades** from which terminal branches, which are effectively end arteries, pass to the wall of the bowel (**6.5.2, 6.5.12**) (the arcades are more numerous in the ileal region of the bowel, so that jejunum can be distinguished from ileum at operation by the 'windows' between the terminal branches, which are tall and narrow in the former and shorter and wider in the latter);

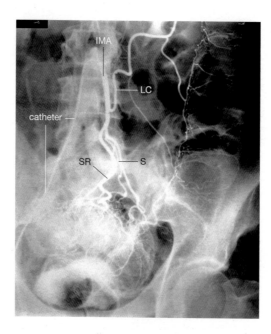

6.5.13 Arteriogram of inferior mesenteric artery (IMA) and its distribution by left colic (LC) sigmoid (S) and superior rectal (SR) branches.

- the **ileo-colic artery**, which passes to the right iliac fossa where it supplies the terminal ileum, the caecum, an **appendicular artery** which passes behind the ileum to reach the appendix via the mesoappendix, and branches to the ascending colon;
- the **right colic artery**, which arises from the right side of the superior mesenteric artery, crosses the posterior abdominal wall immediately beneath the peritoneum, to supply the ascending colon;
- the **middle colic artery**, which arises close to the origin of the superior mesenteric artery and immediately enters the transverse mesocolon to supply the bowel almost up to the splenic flexure.

The **inferior mesenteric artery** (6.5.3, 6.5.13) supplies the hindgut-derived bowel from about the splenic flexure to the upper part of the anal canal. It branches from the aorta opposite the third lumbar vertebra and runs downward and laterally to the left iliac fossa. Its anastomosing branches are:

- the **left colic artery**, which has ascending branches to the splenic flexure and descending branches to the lower descending colon;
- **sigmoid arteries**, which enter the sigmoid mesocolon to reach the bowel wall;
- the **superior rectal artery**, which passes into the pelvis along the line of attachment of the sigmoid mesocolon to reach the rectum. Its branches supply the mucosa of the entire rectum but the muscle of only its upper two-thirds.

Muscle of the middle and lower thirds of the rectum gain arterial supply from the **middle rectal** branch of the internal iliac artery, while the entire anal canal is supplied by the (badly named) **inferior rectal arteries** derived from the internal pudendal branch of the internal iliac artery in the perineum.

The arteries to the colon anastomose to form a 'marginal artery' (6.5.2b, 6.5.3) which is essentially a series of single arcades. This provides an anastomosis of vessels in the mesentery along the length of the large bowel.

The alimentary canal and its derivatives receive about 20% of the cardiac output. Of this, the majority passes through the coeliac and superior mesenteric arteries; the inferior mesenteric artery carries only about 3%.

The arteries supplying both the small and large intestine give numerous terminal branches of supply which pierce the muscular coats of the gut to form a plexus of vessels in its submucous coat (6.5.12). Anastomoses between the branches which enter the bowel are not always plentiful and, if a segment of bowel has to be removed, the viability of the ends to be anastomosed must be assured.

Blood flow through the splanchnic bed is regulated by the autonomic nervous system. Situations that stimulate the sympathetic system cause constriction of the splanchnic (and cutaneous) vascular bed, so that more blood is available to the muscles. When this occurs, the numerous arterio-venous anastomoses in the wall of the upper bowel help to maintain portal vein pressure and liver perfusion. Conversely, during digestion, blood flow to the stomach and small intestine increases considerably; the small intestines receive 50–60% of the superior mesenteric flow at rest, and 60–80% during digestion.

The **veins** of the small and large bowel (6.5.14) follow the arteries with few exceptions, and all drain eventually into the **hepatic portal vein**. By this arrangement the alimentary venous blood passes through the liver which, by absorption and secretion, controls the contents of the blood reaching the systemic circulation.

The **inferior mesenteric vein** does not follow the inferior mesenteric artery but runs upward on the posterior abdominal wall, passing just to the left of the duodeno-jejunal flexure (in the wall of any paraduodenal fossa) to join the splenic vein behind the body of the pancreas. The splenic vein runs to the right behind the pancreas and joins the **superior mesenteric vein** to form the portal vein.

Small anastomotic connections occur between veins draining retroperitoneal parts of the intestines and systemic veins of the body wall. Venous blood from the rectum drains partly into the inferior mesenteric vein and partly into the systemic middle and inferior rectal veins. Both these anastomoses may dilate if portal venous pressure is increased (see p. 163).

Lymphatics draining the intestine and its intrinsic lymphoid tissue (6.5.15) run along its arterial tree to reach the coeliac, superior mesenteric, or inferior mesenteric **preaortic nodes**, lying around the origin of the arteries. Lymph from the rectum drains not only to the inferior mesenteric nodes but also along all the other arteries that supply the rectum.

Qu. 5H *In a patient with a malignant tumour of the rectum, which nodes other than the inferior mesenteric nodes would you need to investigate for secondary deposits of tumour?*

Mesenteric vascular insufficiency

Although there is considerable potential anastomosis between the vessels supplying the bowel, death of an area of bowel due to vascular insufficiency can occur. The capillary circulation of the mucosa of the small intestine is prone to become impaired because:

- portal venous pressure is higher than systemic venous pressure;
- vasomotor reflexes protect other parts of the circulation at the expense of the bowel, and visceral arterioles are very sensitive to circulating catecholamines;
- the vessels supplying the mucosa are constricted by peristaltic contractions;
- the close proximity of the arterial and venous ends of the mucosal capillary loops produces a countercurrent exchange, so that the tips of villi are the first to suffer anoxia.

Arterial disease affects the visceral and systemic circulations alike. The part of the bowel at greatest risk of vascular insufficiency is the splenic flexure of the colon where:

- the faecal material is relatively firm, so that considerable pressure on the mucosal surface is exerted by the peristaltic contractions trying to propel faeces round the acute splenic flexure; and
- the arteries supplying the bowel wall are at the limits of supply of two stem arteries.

Therefore constipation, which allows more water to be removed from faeces making them harder, predisposes to ischaemic damage to the bowel.

Diverticulitis

Small vessels pierce the wall of the colon (**6.5.2b**) to reach the appendices epiploicae and, in doing so, create small points of weakness. If intraluminal pressure is consistently raised (e.g. by chronic constipation) diverticula of the mucosa can herniate through the muscle wall and become inflamed—diverticulitis. (Other diverticula of the bowel are of embryonic origin.)

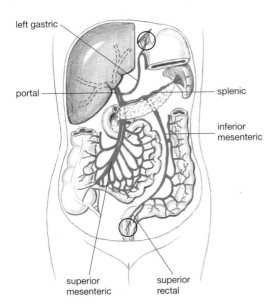

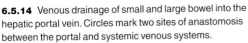

6.5.14 Venous drainage of small and large bowel into the hepatic portal vein. Circles mark two sites of anastomosis between the portal and systemic venous systems.

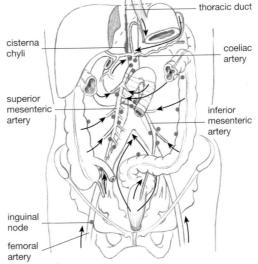

6.5.15 Lymphatic drainage of the alimentary canal. Note that the cisterna chyli receives: an intestinal trunk which drains the intestines and spleen; and right and left trunks which drain the abdominal walls and organs (e.g. kidneys) lying on them.

Lymphatics from the lower part of the anal canal drain outward to inguinal lymph nodes—a reflection of the ectodermal origin of this part of the bowel epithelium.

The lymphatics start as capillaries in the mucosa. In the small bowel villi these receive minute droplets (chylomicrons) of absorbed fats, and are called **lacteals** from the milky appearance of the lymph. The bowel wall also contains free lymphocytes and lymphoid follicles ('gut-associated lymphoid tissue'). In most of the bowel the follicles are small, but larger aggregations (Peyer's patches) are formed in the ileum, and the appendix is particularly richly endowed. Nodes draining lymph from the colon lie in the wall of the bowel, along its mesenteric border, and along the main arteries.

Innervation of the small and large intestines

Before considering this brief summary, you should review the organization of the autonomic nervous system (p. 16).

The innervation of the tract is derived from both the sympathetic and the parasympathetic parts of the system. In addition, large numbers of neurons in the wall of the bowel form an **enteric nervous system**. The efferent fibres modulate peristalsis, sphincter tone, and secretion via the **myenteric plexus** (between the muscle layers) and **submucous plexus** of the enteric nervous system. Many neuroactive substances (e.g. peptides, ATP, nitric oxide) other than noradrenaline and acetylcholine are involved.

- The **sympathetic** innervation controls the splanchnic blood vessels, reduces peristalsis and closes sphincters. It comprises preganglionic and postganglionic neurons. The preganglionic neurons have cell bodies in the lateral horn of the spinal cord (T5–L2) and axons that leave the cord and pass straight through the sympathetic chain to form **splanchnic nerves** which synapse in the coeliac and mesenteric ganglia. The postganglionic neurons in these ganglia have non-myelinated axons which run along the arterial tree to control the vessels and reach the bowel wall.
- The **parasympathetic** innervation stimulates many of the secretions of the alimentary tract. It also has preganglionic and postganglionic neurons. The preganglionic neurons supplying the fore- and midgut are located in a brainstem nucleus of the **vagus nerve**. Its axons are distributed via the coeliac and superior mesenteric ganglia to postganglionic neurons in the wall of the bowel from pharynx to splenic flexure. Preganglionic neurons supplying bowel derived from the hindgut (and other pelvic viscera) are located in the sacral part of the spinal cord (S2–4). Their axons form the **pelvic splanchnic** (pelvic parasympathetic) nerves, which are distributed via the pelvic plexus to bowel from the splenic flexure to the rectum. Cell bodies of the postganglionic fibres supplying the bowel are located in the enteric plexus.

Visceral *sensory* fibres from the bowel run with the efferent fibres of both the sympathetic and parasympathetic systems. Many of these fibres form the afferent loop of alimentary reflexes, sensing distension and gut contents; and pass largely along parasympathetic pathways, especially the vagus.

Fibres mediating the sensation of pain from the viscera enter the central nervous system largely at spinal levels. Visceral pain is not appreciated in the position of the adult viscera, but is 'referred' to the skin supplied by the spinal cord segment at which the visceral afferents first synapse in the cord. Referred pain is not sharply localized and is usually felt maximally in the midline. From foregut derivatives it is felt in the epigastrium; from midgut derivatives in the umbilical region, and from hindgut derivatives in the lower abdomen.

Qu. 5l *If an appendix became inflamed where would you expect the pain to be felt initially? If the inflamed appendix touched the parietal peritoneum of the right iliac fossa, how would the quality and distribution of the pain change, and why?*

Bowel functions are not dependent on an intact extrinsic innervation. If the sympathetic and/or parasympathetic fibres are damaged, peristalsis and secretions cease for a while, but then re-establish themselves through the intrinsic activity of the enteric nervous system. However, the system can no longer respond to stimuli outside the alimentary tract which would normally modulate its responses.

Bowel function is also co-ordinated by a system of **enteroendocrine cells** in the mucosa. These respond both to the contents of the bowel and to its innervation.

Computerized imaging

CT and MRI (**6.5.16** and see Section 6.11) examination shows both the small and large intestine parts of the bowel clearly. Small intestine has fluid content and does not normally contain gas. Large intestine is distinguished both by its position and by the presence of irregular faecal material mixed with gas. The sigmoid colon stores faeces, but the rectum is normally empty of faeces except in the period prior to defaecation.

Ultrasound is also used to examine the wall of the rectum and the adjoining tissues (**6.5.17**).

Living anatomy

Visual examination of the anterior abdominal wall usually reveals nothing about the underlying bowel

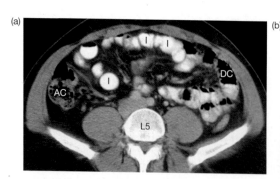

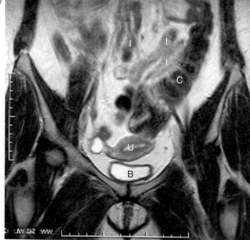

6.5.16 (a) Axial CT of abdomen through L5; (b) coronal MRI of female pelvis. Small intestine (I), ascending colon (AC), descending colon (DC), distal sigmoid colon (C) leading to rectum; uterus (U), bladder (B).

and only parts of the large bowel in which the faeces are sufficiently solid can be identified by palpation. Thus, the lower part of the descending colon can often be rolled beneath the fingers in the left iliac fossa.

Percussion reveals areas of resonance created by gas—again often over the descending colon. Bowel movements are not visible unless abnormally increased by obstruction. However, they create sounds that can readily be heard with a stethoscope. An absence of bowel sounds suggests that peristaltic move-ments have ceased. If they are very loud, there may be obstruction.

On yourself or your partner draw the transpyloric plane, the transtubercular plane, the two lateral lines, and the surface markings of the pylorus, duodenum and duodeno-jejunal flexure, the root of the mesentery, and the different parts of the colon, including the root of the appendix (McBurney's point) (**6.5.18**).

Manual clinical examination of the lowest part of the rectum and anal canal is performed with a gloved index finger inserted gently though the lubricated anus. It is used to detect changes in the anal canal, rectum, and adjacent organs (see pp. 186, 197, 210).

Questions and answers

Qu. 5A *How could you use the mesentery of the small bowel to determine, at operation, which is the proximal end of a loop of bowel which presented through an incision?*

Answer Run a finger down the mesentery until its attachment to the posterior abdominal wall is reached (untwisting the small bowel as necessary). Then, with the mesentery straightened, the cranial end with respect to the attachment must be the proximal end of the loop. (Normal peristalsis may not be present.)

Qu. 5B *What is a Meckel's diverticulum a remnant of?*

Answer The vitello-intestinal duct which normally disappears, but occasionally remains as a remnant about 60 cm from the ileo-caecal junction.

Qu. 5C *On which muscles and what organs does the ascending colon lie?*

Answer The ascending colon lies on the posterior abdominal wall on iliacus, then quadratus lumborum and transversus abdominis. It is in contact with the lateral part of the lower pole of the right kidney and its hepatic flexure lies beneath the right lobe of the liver, from which it is separated by the peritoneal cavity.

Qu. 5D *From which part of the primitive gut tube is the descending colon derived, and therefore which artery supplies it?*

Answer The descending colon is derived from hindgut and is therefore supplied by the inferior mesenteric artery.

Qu. 5E *Where might secondary deposits from cancer of the colon be found?*

Answer In the liver, in local lymph nodes, in lymph nodes along the superior or inferior mesenteric artery, or in adjacent tissues if the tumour has invaded through the wall of the colon.

Qu. 5F *How does the appearance of the colon differ from that of the small bowel when filled with contrast medium?*

Answer The haustrations and smooth mucosa of the colon produce an appearance very different from the fine stippled appearance produced by the circular folds and villi in the small bowel.

Qu. 5G *What is the reason for this triangular appearance of the transverse colon?*

Answer The triangular endoscopic appearance of the transverse colon is produced by the attachments of the taenia coli.

Qu 5H *In a patient with a malignant tumour of the rectum, which nodes other than the inferior mesenteric nodes would you need to investigate for secondary deposits of tumour?*

Answer The rectum receives arterial blood from the superior, middle and inferior rectal arteries and drains into the corresponding veins. Lymphatic drainage is therefore into lymph glands around the superior rectal artery and also around the internal iliac, inferior rectal and pudendal arteries.

Qu. 5I *If an appendix became inflamed where would you expect the pain to be felt initially? If the inflamed appendix touched the parietal peritoneum of the right iliac fossa, how would the quality and distribution of the pain change, and why?*

Answer Appendicitis pain presents first as poorly localized central abdominal pain ('referred' due to stimulation of enteric nociceptors). If the parietal peritoneum becomes irritated the pain becomes sharply localized in the region of the inflamed peritoneum—usually the right iliac fossa.

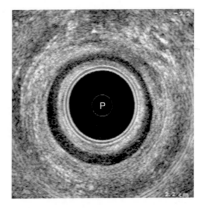

6.5.17 Ultrasound of rectum, showing the different layers of muscle. The innermost white circle is the surface of the ultrasound probe (P).

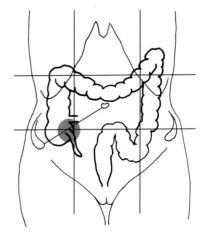

6.5.18 Surface markings of the large bowel. Note the position of the root of the appendix one-third of the way along a line from the anterior superior iliac spine to the umbilicus. The body of the appendix varies greatly in position.

Liver and biliary tract, pancreas, and spleen

Liver and biliary tract, pancreas, and spleen

The liver is the largest organ in the body. It receives oxygenated blood from the hepatic artery and also nutrient-rich blood, derived from the gastrointestinal tract, which reaches it via the hepatic portal vein. It has several life-sustaining functions. These include the production of many plasma proteins, including albumin; a central role in the intermediary metabolism of carbohydrates, lipid, and proteins; and the storage of glucose as glycogen. In addition, its macrophages remove particulate matter, including effete red cells. The bile produced by the hepatocytes conveys into the duodenum pigmented waste products derived from red cell breakdown and bile salts important in the emulsification of fats. Bile is also an important secretion pathway for cholesterol.

When bile is not required in the duodenum it is stored and concentrated in the gall bladder. Before joining the duodenum, the bile duct joins the main pancreatic duct, which carries the exocrine secretions of the pancreas (digestive enzymes, ions) to the duodenum. The islets of Langerhans embedded in the pancreas are endocrine tissue with a primary role in the control of nutrient metabolism.

The spleen is part of the reticuloendothelial system; it filters the blood and produces antibodies.

Development

During the fourth week of fetal life, while the rotation of the midgut is occurring, a **hepato-pancreatic diverticulum** grows out from the primitive endodermal gut tube at the junction of the foregut and midgut (**6.6.1**). The **hepatic diverticulum** grows into the mesoderm of the septum transversum and soon divides into **cystic** and **hepatic** parts. The hepatic part divides into right and left branches which, with mesodermal components derived

from the septum transversum, form the right and left halves of the liver and their biliary drainage ducts.

Within the growing hepatic tissue, vascular sinusoids develop. These connect with the right and left umbilical and vitelline veins. These veins run through the septum transversum to the sinus venosus at the caudal end of the heart tube, anastomosing *en route* with hepatic sinusoids. The right umbilical vein regresses and the left umbilical vein becomes continuous with a preferential

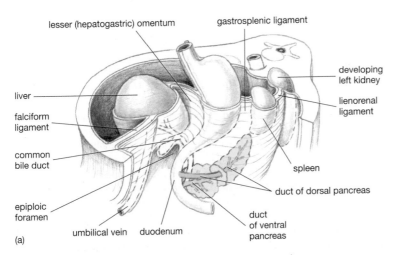

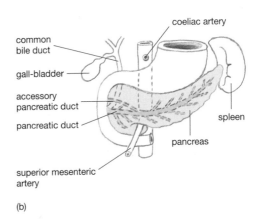

6.6.1 (a) Development of liver and biliary tract and (b) adult arrangement.

channel through the sinusoids—the **ductus venosus**—which drains into the right horn of the sinus venosus via the fused cranial ends of the right umbilical and vitelline veins. By this means, most of the oxygen- and nutrient-rich blood from the placenta passes directly through the fetal liver on its way to the heart. After birth the left umbilical vein and ductus venosus both close, leaving fibrous remnants, the **ligamentum teres** and **ligamentum venosum.**

The vitelline veins also anastomose with the liver sinusoids, but not with the ductus venosus. They become the system of veins which drain the developing gut and form the hepatic portal vein so that, after birth, nutrient-rich alimentary venous blood passes to the liver before reaching the systemic circulation.

At the cranial end of the septum transversum, the hepatic sinusoids drain into the fused right umbilical and vitelline veins. Together these form the right hepato-cardiac channel, which will become the part of the inferior vena cava between the liver and the heart. The cranial ends of the left umbilical and vitelline veins regress, as does the left horn of the sinus venosus into which they originally drained (p. 69).

The **cystic diverticulum** that will form the gall bladder and cystic duct develops from the ventral edge of the hepatic diverticulum before its division into right and left branches (see **6.4.1**).

The pancreatic part of the hepato-pancreatic diverticulum forms a **ventral pancreatic rudiment**; a larger **dorsal pancreatic rudiment** develops more cranially from the dorsal margin of the gut tube. The dorsal pancreas grows into the dorsal mesogastrium to form the upper part of the head of the definitive pancreas and its neck, body, and tail. At the same time, differential growth of the duodenum carries the developing biliary tree and ventral pancreas dorsally and to the left with the ventral mesogastrium which links the foregut to the developing liver. The ventral pancreas, which forms the lower part of the head and the uncinate process of the definitive pancreas, thus comes to lie immediately below, and then fuses with, the larger dorsal pancreas. As the dorsal and ventral rudiments come together, the superior mesenteric artery and vein are trapped between them.

This relative migration of the biliary tree, coupled with the fusion of the duodenal loop to the right side of the posterior abdominal wall, causes the bile duct and portal vein to lie behind the first part of the duodenum.

Failure of normal development can result in pancreatic tissue forming all round the duodenum to become an **annular pancreas**, which may constrict the duodenum.

Within the pancreas, the duct systems of its dorsal and ventral components are initially independent. Later, the duct of the ventral pancreas takes over the drainage of most of the dorsal pancreas to form the principal pancreatic duct; the original duct of the dorsal pancreas then opens into the duodenum as the accessory pancreatic duct.

Migration of the attachment of the dorsal mesogastrium to the left, and fusion of its lower part to the posterior abdominal wall as far as the lower border of the pancreas, means that the majority of the pancreas lies retroperitoneally on the posterior abdominal wall. However, the tail of the pancreas extends laterally into the part of the dorsal mesogastrium which forms the lienorenal ligament, and thus extends as far as the hilus of the spleen.

The **spleen** develops in the dorsal mesogastrium just beyond the tip of the dorsal pancreas (**6.6.1** and see **6.4.1**). Small accessory spleens may form in the region, and occasionally the organ retains its lobulated fetal form.

During development, both the liver and the spleen are sites of haemopoiesis and are proportionally larger in the fetus than in the adult.

Liver

Form and position

The liver (**6.6.2**) is the largest single organ in the body. It has a pliable and firm but friable parenchyma held within a thin fibrous capsule, much of which is covered by peritoneum. It is wedge-shaped, with a large upper right surface lying against the right dome of the diaphragm and lower ribs, and a narrow apex which extends to the left and ends next to the fundus of the stomach. Its bulk is largely hidden behind and protected by the right lower ribs, occupying nearly all the right hypochondrium and extending left into the epigastrium and left hypochondrium (see **6.6.11**).

Qu. 6A *What might be the consequences for the liver if lower ribs on the right are fractured?*

Functional and anatomical divisions

Functionally, the liver comprises a right and left half (each with four segments; **6.6.3**) which are more or less fused together. Each segment has its own relatively separate vasculature and biliary drainage, an important consideration in liver surgery. However, because the undersurface of the liver presents four distinct areas (**6.6.2b**) four anatomical 'lobes' have been described.

The larger **right lobe** (segments V–VIII) extends on the inferior surface as far as the **porta hepatis**—the hilus of the liver through which the hepatic artery, portal vein, and hepatic ducts enter and leave the liver; on the anterior surface (**6.6.2a**), the anatomical right lobe extends as far as the attachment of the falciform ligament. Segments II and III form the anatomical left lobe, but functionally the left half of the liver includes segment IV (the quadrate lobe, situated between the gall bladder and the fissure for the falciform ligament on the undersurface of the liver) and segment I, the **caudate lobe**, which lies in the roof of the lesser sac between the inferior vena cava and a fissure containing the ligamentum venosum.

The caudate lobe is continuous with the right lobe of the liver via the caudate process, a narrow strip of liver tissue between the porta hepatis and the inferior vena cava in the roof of the epiploic foramen. The fissure marks the course of the fetal ductus venosus. It extends from the ligamentum teres in the falciform ligament at the left extremity of the porta hepatis to the upper end of the inferior vena cava.

Supports and peritoneal attachments

The liver is the heaviest organ within the abdominal cavity. It is supported by:

- **intra-abdominal pressure**, which is created largely by the tone of the abdominal wall muscles;
- the **hepatic veins**, which open almost directly from the substance of the liver into the inferior vena cava.
- **peritoneal attachments**.

Most of the surface of the liver (and gall bladder) is covered with peritoneum. This allows considerable movement with diaphragmatic respiration. Peritoneum covering the anterior surface of the liver is reflected on to the anterior abdominal wall above the umbilicus to form the sickle-shaped **falciform ligament**, in the lower free edge of which lies the ligamentum teres.

Peritoneum covering the upper surface of the liver is reflected on to the undersurface of the diaphragm, leaving the liver capsule in contact with the diaphragm (**bare area**). The bare area of the left lobe is small, but that of the right lobe is much larger, and includes the area in which the inferior vena cava is embedded and the area in contact with the right suprarenal gland.

The reflections of peritoneum on to the diaphragm and posterior wall are called **coronary ligaments**; their narrow lateral extremities form the **right** and **left triangular ligament**.

Extending from the lesser curvature of the stomach to the porta hepatis and fissure for the ligamentum venosum is the **lesser omentum**. This separates the lesser sac (behind) from the greater sac of the peritoneum (in front). The caudate lobe of the liver projects into the roof of the lesser sac.

In the free right border of the lesser omentum are the common hepatic and common bile duct, with the hepatic artery on its left, and the hepatic portal vein posteriorly. Immediately behind the free border is the epiploic foramen.

In a plain radiograph the liver forms an opacity in the right hypochondrium. Its upper border is clearly shown by the contrast with gas in the lungs above (see **5.3.10**);

Liver biopsy

It may be necessary for diagnostic purposes to take a biopsy specimen from the liver. This is done by inserting a large-bore needle into the liver. If the liver problem is focal, the needle can be guided by CT or ultrasound; if generalized, the needle is inserted through an area of maximal liver dullness in the right mid-axillary line. The needle will pass through chest wall, the costo-diaphragmatic recess of the pleural cavity, the diaphragm, and probably also the peritoneal cavity before entering the liver.

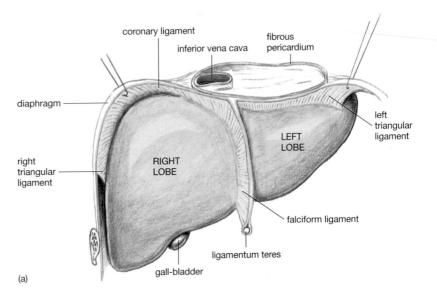

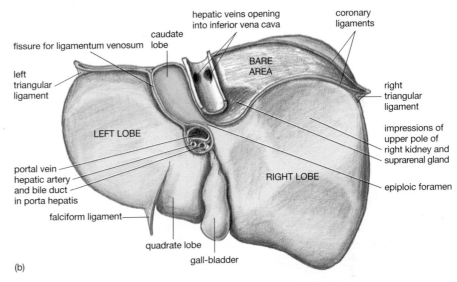

6.6.2 (a) Anterior and (b) posterior views of the liver and its peritoneal reflections.

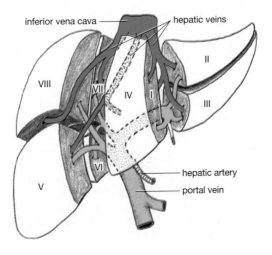

6.6.3 Functional segments of the liver (modified from the *Oxford Textbook of Surgery*).

its lower border is not distinct unless gas in adjacent parts of the colon provides a contrast.

Biliary system

The parenchymal cells of the liver continuously secrete bile into microscopic **canaliculi** formed by and lying between the cells. The canaliculi drain into intrahepatic tributaries of the bile ducts which unite to form **right** and **left hepatic ducts** at the porta hepatis. These join to form the **common hepatic duct**, which enters the upper end of the lesser omentum. Here it is joined by the **cystic duct** of the **gall bladder** to form the **common bile duct**, which transports bile to the duodenum (**6.6.4**). Bile is allowed to enter the duodenum only intermittently, largely after meals. Between meals the gall bladder acts as a store of bile.

The **gall bladder** is an epithelium-lined sac with a smooth muscle wall. (In a cadaver it is stained green by bile which penetrates its wall after death.) It is usually firmly attached to the undersurface of the liver, although it may have a small mesentery. It is described as having a **fundus** (the expanded end which hangs just below the costal margin), a **body**, and a **neck**. The body and fundus hang down from the neck, so that the fundus lies immediately in front of the superior duodenal flexure; the narrow neck is continuous with the **cystic duct**. In the neck and duct the mucous membrane forms ridges which create a **spiral valve**.

Qu. 6B *Why do gallstones tend to collect in the gall bladder rather than passing down the biliary tract?*

The **common bile duct** commences anterior to the portal vein in the free margin of the lesser omentum. It passes down in front of the opening to the lesser sac, then behind the first part of the duodenum; it then passes through the substance of the head of the pancreas to join with the main pancreatic duct. The conjoined ducts then penetrate the medial wall of the second part of the duodenum obliquely, at about its midpoint. The opening into the duodenum (see **6.6.5**) is characterized by a protrusion—the **duodenal papilla**—which denotes the site of a strong **sphincter** (sphincter of Oddi).

Qu. 6C *What is the significance of a sphincter at this site?*

The biliary system can be visualized radiologically only if it contains radiopaque material. This can be introduced through a cannula introduced into the biliary system during an operation, or passed via an endoscope through the duodenal papilla (cholangiography), or alternatively by intravenous administration of a contrast agent that is excreted in the bile (cholecystography). The endoscopic approach can also be used to produce endoscopic retrograde cholangiopancreatograms (ERCPs) (**6.6.6, 6.6.7**) which can show both the biliary and pancreatic ducts.

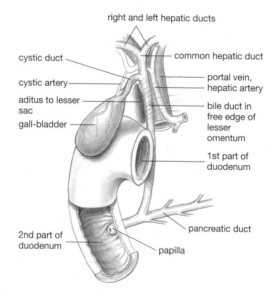

right and left hepatic ducts

cystic duct

common hepatic duct

cystic artery

portal vein, hepatic artery

aditus to lesser sac

bile duct in free edge of lesser omentum

gall-bladder

1st part of duodenum

2nd part of duodenum

pancreatic duct

papilla

6.6.4 Biliary system and blood supply of the gall bladder.

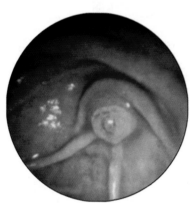

6.6.5 Endoscopic view of the duodenal papilla.

6.6.6 ERCP. The cannula has been passed through the duodenal papilla and into the common bile duct. The spiral valve of the cystic duct, tributaries of the hepatic ducts within the liver, and tortuous main pancreatic duct can also be seen. In this and in **6.6.7** the 'scotty dog' appearance of the lumbar spine shows that the view is oblique. The endoscope used to introduce the cannula is obvious.

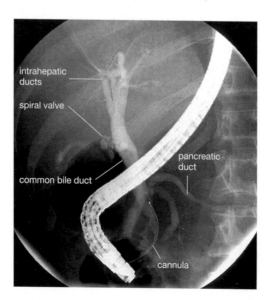

intrahepatic ducts

spiral valve

pancreatic duct

common bile duct

cannula

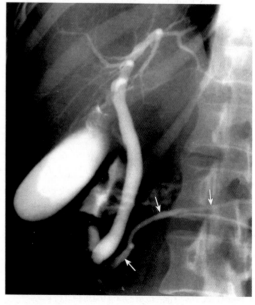

6.6.7 ERCP. Biliary and pancreatic (*arrowed*) duct systems.

Gallstones and jaundice

Gallstones (**6.6.8**) may form in the bile within the gall bladder, where the stored bile is concentrated by removal of water and electrolytes. The stones are usually composed of cholesterol and/or bile pigments and may become calcified. They form slowly and vary greatly, from a single large stone to multiple small stones.

One of the functions of the liver is the breakdown of the haem pigment of aged red cells and the excretion of the pigmented products in the bile. If the products cannot be excreted (or if their production exceeds the rate of excretion, for example because of haemolysis) then **jaundice** occurs.

The gall bladder may become inflamed (cholecystitis). This is usually associated with the presence of gallstones, perhaps with added infection, and surgical removal of the gall bladder (cholecystectomy) is often required. This operation is now often done by a laparoscopic approach.

Pain from an infected or obstructed gall bladder is felt in the epigastrium if the parietal peritoneum is not involved. If parietal peritoneum is irritated, the pain is felt over the gall bladder and may be referred to the right shoulder tip.

An inflamed gall bladder can become adherent to the duodenum (this can inhibit peristalsis—gallstone ileus) and gallstones may even erode into the duodenum.

Ultrasound imaging (**6.6.8**) is now frequently used to examine the liver, gall bladder, and the bile ducts, as is CT and MRI (**6.6.10**).

Qu. 6D *Would jaundice result after blockage of (a) the cystic duct; (b) the right hepatic duct; (c) the common bile duct?*

Qu. 6E *If a gallstone became lodged at the sphincter of Oddi, which other organ would be affected?*

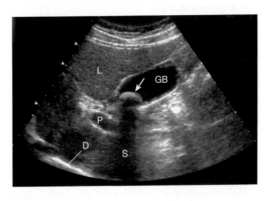

6.6.8 Ultrasound of gall bladder (GB) containing a gallstone (*arrow*) which casts a shadow (S). The patient is supine; diaphragm (D), liver (L), portal vein (P).

Blood supply and lymphatic drainage of the liver and biliary system

The liver receives two separate blood supplies (**6.6.3**):
- A **systemic** supply from the **hepatic artery**, which arises from the coeliac axis of the abdominal aorta; this contributes 20% of the total supply at systemic pressure (80–120 mmHg). The hepatic artery passes forwards from the aorta, beneath the epiploic foramen, to reach the first part of the duodenum; it then runs upward in the free edge of the lesser omentum and terminates in the porta hepatis by dividing into right and left hepatic branches which supply the two functional halves of the liver. The gall bladder and cystic duct are supplied by a small **cystic artery** given off the right hepatic artery; the lower part of the common bile duct is supplied from the gastroduodenal artery.
- The **hepatic portal vein**, which carries nutrient- and hormone-rich blood from the abdominal gut, pancreas and spleen. The portal vein provides 80% of the total supply to the liver at a low pressure (5–10 mmHg). It has no valves. The portal vein is formed by the junction of the splenic and superior mesenteric veins behind the neck of the pancreas. It runs upward behind the first part of the duodenum and then in the lesser omentum to reach the porta hepatis where it divides into right and left branches.

Qu. 6F *Within the lesser omentum, which vessel lies closest to the free margin?*

Portal hypertension

If the liver substance becomes fibrosed as a result of repeated inflammation or damage by alcohol—cirrhosis—the low-pressure perfusion of the liver with portal venous blood is impeded and portal venous pressure rises (portal hypertension). When this occurs, the veins in the oesophagus which form anastomoses between the portal and systemic venous systems enlarge and become varicose (oesophageal varices; see **6.4.6b**). In addition, excess fluid collects in the peritoneal cavity (ascites).

It might be expected that the anastomosis in the rectum would also often become varicose, but this is much less common. Most dilated rectal veins (haemorrhoids) have a different origin (p. 207).

Bleeding from oesophageal varices can be life-threatening. To relieve portal hypertension in such cases, an anastomosis can be created surgically between the portal vein and the inferior vena cava, or between the splenic vein and left renal vein. A new method (transjugular intrahepatic portosystemic shunt; TIPS) which avoids a major abdominal operation, creates the anastomosis within the liver. A fine instrument is inserted into the internal jugular vein and passed down through the superior vena cava, right atrium and inferior vena cava into a hepatic vein. A needle is then passed into the substance of the liver to enter a branch of the portal vein. This then acts as a guide for a wider metal tube (stent) which creates an anastomosis between the portal and systemic venous circulations.

Anastomoses between portal and systemic veins

Venous blood from the abdominal alimentary tract drains to the liver via the hepatic portal vein. The remainder of the body drains to the systemic circulation. Where the two are in continuity or contact, there are anastomoses between the two systems.

Qu. 6G *What are the two major sites of anastomosis between the portal and systemic venous systems?*

Other sites of anastomosis are:

- the umbilicus. Small veins running with the ligamentum teres anastomose with veins of the anterior abdominal wall. When varicose, these form a 'Medusa's head' (caput medusae) of distended veins radiating from the umbilicus;
- at the attachments of retroperitoneal bowel. Veins of the bowel anastomose with those of the posterior abdominal wall.

Venous drainage

Venous blood from the liver drains through three very short, wide veins which open directly into the inferior vena cava as it passes through the posterior aspect of the liver. In severe abdominal trauma these veins can be ripped from the inferior vena cava, usually with fatal results.

The gall bladder drains by a cystic vein into the right branch of the portal vein.

Lymphatic drainage

Most liver-derived lymph, which is very rich in protein by reason of the permeable nature of the liver vascular sinusoids, passes either to nodes around the upper end of the inferior vena cava or to nodes at the porta hepatis. Lymph from the biliary tree ends in nodes at the porta hepatis and along the common bile duct.

Innervation of the liver and biliary system

The liver receives both sympathetic and parasympathetic fibres which enter at the porta hepatis. Their functions are not well understood. In the past, when surgeons treated peptic ulceration by interrupting vagal innervation to the acid-producing area, they often tried to divide only the branches to the body of the stomach, thus preserving fibres to the liver and other parts of the alimentary tract.

The gall bladder receives largely sympathetic and sensory nerves.

Qu. 6I *What causes contraction of the gall bladder after a meal?*

Impaction of a gallstone in the biliary tree, or infection of the biliary system can cause severe **pain**. If the biliary system is blocked, colicky pain is felt in the epigastrium and radiates through to the back. If the gall bladder is infected, inflammation spreads to the overlying parietal peritoneum so that the pain is more sharply localized in the right hypochondrium and may also be referred via the phrenic nerve (C4) which supplies the diaphragmatic peritoneum.

Qu. 6J *Where might pain referred via the phrenic nerve be felt?*

The biliary tree may be compressed and blocked by tumours, either primary or secondary, affecting the nodes at the porta hepatis or the head of the pancreas. Such obstruction does not necessarily cause pain.

Pancreas

The **pancreas** (**6.6.9**) is a soft, finely lobulated gland which runs almost transversely across the posterior abdominal wall from the duodenum to the spleen. It is unusual in that it does not have a distinct fibrous capsule. It comprises a head with an uncinate process, a neck, body, and a tail. The **head** lies in the concavity of the duodenum in front of the inferior vena cava; its **uncinate process** extends leftward from the head, separated from the neck by the superior mesenteric vessels; the **neck** lies on the portal vein and common bile duct immediately below the pylorus; the **body** crosses in front of the aorta, the left crus of the diaphragm, the left suprarenal gland, and left kidney; the **tail** runs in the lienorenal ligament to reach the spleen.

The transverse mesocolon is attached along the lower border of the body and neck and across the head of the pancreas (see **6.5.3**). Because the spine, aorta, and inferior vena cava project markedly in the midline of the posterior abdominal wall, the pancreas is curved around this, so that its body passes obliquely posteriorly to reach the spleen. This is very obvious on CT (see **6.3.11** and **5.6.11**) and MRI (**6.6.10**).

6.6.9 Pancreas and its immediate relationships.

Qu. 6K *Which peritoneal space and which organ lie in front of the upper part of the body of the pancreas?*

The pancreas has both **endocrine** and **exocrine** components:

- The **exocrine** tissue is composed of compound tubulo-alveolar glands which secrete digestive enzymes and alkali into a series of ducts. These ducts open into a **principal pancreatic duct** which runs through the length of the tail and body, then downward through the head to join the common bile duct and drain into the duodenum (**6.6.6, 6.6.7**). An accessory duct is usually present, draining the upper part of the head.
- The **endocrine** tissue forms the **islets of Langerhans**; these produce two principal hormones, insulin and glucagon, which, on secretion into the bloodstream, exert profound influences on carbohydrate metabolism and on growth. Somatostatin and pancreatic polypeptide are also produced by islets and appear to have mainly local effects.

Spleen

The spleen, like the liver, consists of a highly vascular pulp surrounded by a thin fibrous capsule and a layer of visceral peritoneum. It lies in the left hypochondrium beneath the ninth to eleventh ribs and is variable in shape, although its antero-posterior border is usually marked by a notch. With ageing it atrophies to about one-third its size in a young adult. It has a convex outer surface which lies against the diaphragm, and a concave inner surface which makes contacts with the stomach, kidney, and splenic flexure of the colon. The tail of the pancreas abuts the spleen at its vascular hilus.

Qu. 6L *How much of the spleen is surrounded by peritoneum and by what is it connected to the posterior abdominal wall and stomach?*

The vascular splenic **red pulp** acts as a filter and its macrophages remove effete blood cells and particulate matter. In some animals, but not in humans, the spleen acts as an effective reservoir of red cells which can be ejected into the circulation by contraction of smooth muscle in the trabeculae and capsule.

The pulp also contains numerous lymphoid follicles, aggregated as **white pulp**. With the macrophages these follicles form part of the antibody-producing system. The spleen can become markedly enlarged in infections such as malaria, or when involved by tumours of haemopoietic (blood-forming) cells. In such cases its lower extremity and notched anterior border may become palpable below the costal margin.

Blood supply and lymphatic drainage of the pancreas and spleen

The head of the pancreas receives its arterial supply from **superior** and **inferior pancreatico-duodenal arteries** derived, respectively, from the coeliac and superior mesenteric arteries (**6.6.9**, see **6.4.14**). The rest of the pancreas is supplied by the large **splenic artery**, a main branch of the coeliac artery which follows an undulating course along the upper part of the body and tail of the pancreas, which it supplies. Venous blood from the pancreas drains into corresponding veins to reach the portal vein, which is formed behind the neck of the pancreas.

When the splenic artery reaches the hilus of the spleen, it divides into a number of branches. Some enter the spleen; others continue to supply the stomach (short gastric and left gastro-epiploic arteries).

The splenic vein carries splenic venous blood to the hepatic portal vein and is joined by the inferior mesenteric vein behind the body of the pancreas.

Qu. 6M *What changes would you expect to find in the spleen of a person with marked portal hypertension?*

Lymphatic vessels of the pancreas and spleen follow the course of the main blood vessels to drain into lymph glands situated around the coeliac axis.

Innervation of pancreas and spleen

Sympathetic and parasympathetic nerves from the coeliac plexus reach the pancreas. Sympathetic fibres influence primarily its vascular bed, but also mediate pain, which is felt in the epigastrium and often radiates to the back. Vagal parasympathetic fibres stimulate secretion of both pancreatic enzymes and insulin. Sympathetic nerves to the spleen supply the blood vessels and the smooth muscle of the capsule and trabeculae.

Computerized imaging

The liver, gall bladder, pancreas and spleen are all well visualized by MRI (**6.6.10**) and CT (see **5.6.1l, 6.3.11, 6.13.1b, 7.2a**).

Living anatomy

Percussion of the abdomen and lower chest wall will reveal the area of dullness created by the presence of the liver. In a thin subject you may just be able to feel

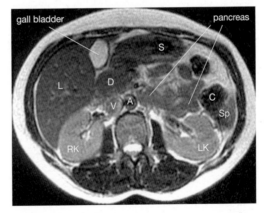

6.6.10 MRI showing gall bladder and pancreas. Aorta (A), splenic flexure of colon (C), duodenum (D), right and left kidney (RK, LK), liver (L), spleen (Sp), inferior vena cava (V).

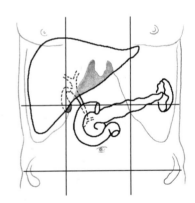

6.6.11 Surface marking of duodenum, pancreas, liver, biliary system, and spleen.

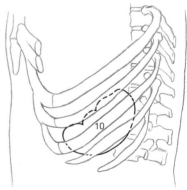

6.6.12 Surface marking of spleen in relation to ninth to eleventh ribs on the left side.

the soft lower border of the liver emerging beneath the costal margin on a deep inspiration.

The spleen lies deeply in the left hypochondrium; its anterior border will not be palpable unless it has at least doubled in size.

On yourself or a partner, draw the outline of the liver, biliary tree, duodenum, pancreas and spleen (6.6.11, 6.6.12), noting the anatomical landmarks that define the normal position of each. Then add the aorta, hepatic and splenic arteries, and superior mesenteric artery. In this way you will build up a good mental picture of the upper abdomen.

Questions and answers

Qu. 6A *What might be the consequences for the liver if lower ribs on the right are fractured?*

Answer The fractured ribs might pierce the diaphragm and pierce the liver, causing intraperitoneal bleeding and shoulder-tip pain.

Qu. 6B *Why do gallstones tend to collect in the gall bladder rather than passing down the biliary tract?*

Answer Gallstones form in the bile within the gall bladder where bile is concentrated by removal of water and electrolytes.

Qu. 6C *What is the significance of a sphincter at this site?*

Answer The sphincter of Oddi surrounds the conjoined common bile duct and main pancreatic duct as they enter into the second part of the duodenum. When acidic fluids from the stomach enter the duodenum the sphincter of Oddi contracts to prevent the entry of this fluid into the pancreas, where it would cause damage.

Qu. 6D *Would jaundice result after blockage of (a) the cystic duct; (b) the right hepatic duct; (c) the common bile duct?*

Answer (a) No; (b) no, provided the left duct was patent; (c) yes.

Qu. 6E *If a gallstone became lodged at the sphincter of Oddi, which other organ would be affected?*

Answer The pancreas would also be obstructed by a stone obstructing the sphincter of Oddi.

Qu. 6F *Within the lesser omentum, which vessel lies closest to the free margin?*

Answer The common bile duct.

Qu. 6G *What are the two major sites of anastomosis between the portal and systemic venous systems?*

Answer Portal systemic anastomoses consistently occur at the lower end of the oesophagus (between left gastric and oesophageal veins) and at the recto-anal junction (between superior and middle rectal veins).

Qu. 6H *At what site are the hepatic portal vein and inferior vena cava closely apposed, and what separates them?*

Answer The portal vein and inferior vena cava lie adjacent at the epiploic foramen separated only by the aditus to the lesser sac.

Qu. 6I *What causes contraction of the gall bladder after a meal?*

Answer Gall bladder contraction is caused by the hormone cholecystokinin (= pancreozymin) secreted by the duodenum in response to a meal, especially one with a high fat content.

Qu. 6J *Where might pain referred via the phrenic nerve be felt?*

Answer Pain referred from the phrenic nerve can be felt in the distribution of C4, especially over the right shoulder.

Qu. 6K *Which peritoneal space and which organ lie in front of the upper part of the body of the pancreas?*

Answer The lower recess of the lesser sac of the peritoneum and the stomach lie in front of the upper part of the body of the pancreas; the transverse mesocolon is attached along the lower border of the body of the pancreas.

Qu. 6L *How much of the spleen is surrounded by peritoneum and by what is it connected to the posterior abdominal wall and stomach?*

Answer All but the hilus of the spleen is surrounded by peritoneum. At the margins of the hilus the spleen is connected to the posterior abdominal wall by the lienorenal 'ligament' and to the greater curvature of the stomach by the gastrosplenic 'ligament'.

Qu. 6M *What changes would you expect to find in the spleen of a person with marked portal hypertension?*

Answer The spleen would be enlarged and engorged with blood as a result of portal hypertension.

Urinary system

Urinary system

The urinary system comprises the two **kidneys** and their associated drainage systems. The kidneys filter the blood and control the excretion of waste products and the conservation of essential ions and water. They therefore have a very profuse blood supply. Urine produced by kidneys is conveyed by the **ureters** to the **bladder**, which acts as a temporary store for the urine. Urine is voided from the bladder through the **urethra**. In women the urethra is short and opens into the vestibule of the perineum; its function is simply as a conduit for urine. In the male the urethra is much longer and is used by both the urinary and reproductive systems. The proximal part of the male urethra is surrounded by the prostate gland which, like the seminal vesicles and vasa deferentia, all discharge their contents into it. The distal part of the male urethra continues through the length of the penis. In both sexes, the voiding of urine (micturition) is stimulated by parasympathetic nerves to the bladder, and controlled by two sphincters, an inner involuntary sphincter situated at the neck of the bladder, and an outer voluntary sphincter surrounding the urethra just before it pierces the perineal membrane.

Development (6.7.1)

In all vertebrates the urinary excretory organs develop in longitudinal columns of **intermediate mesoderm** which form on either side of the midline in association with branches of the aorta. In mammals the kidneys develop from the most caudal part of the intermediate mesoderm—the **metanephros**—within the future pelvis, but subsequently ascend from the pelvis to the posterior abdominal wall (**6.7.1**).

A ureteric (metanephric) bud grows out from the caudal end of the mesonephric duct which runs down longitudinally through the intermediate mesoderm to the cloaca. The ureteric bud branches repeatedly and each branch induces in the surrounding metanephric mesoderm the development of a nephric tubule with which it will fuse. Each nephric tubule, with a blood vessel derived from the primitive aorta, will form the part of a nephron that extends from the glomerulus to the collecting tubule; the remainder of the urinary tract from the collecting tubules to the bladder is formed from the metanephric (ureteric) duct. For the formation of the urethra, see p. 205.

The cloaca originally receives both the mesonephric duct and the hind gut, but downward growth of the urorectal septum separates off an anterior urogenital sinus which forms most of the bladder.

The caudal ends of the mesonephric duct and its ureteric bud become incorporated into the wall of the developing bladder. During this process the two rotate so that the ureters enter the upper lateral aspects of the bladder above the mesonephric ducts, while the mesonephric ducts enter the part of the urogenital sinus that will narrow to form the proximal part of the urethra. Incorporation of the distal ends of the mesonephric ducts and their ureteric buds into the developing bladder gives rise to an area of the bladder wall—the trigone of the bladder—situated between the ureteric orifices and bladder neck.

The bladder, therefore, develops from three components:
- the urinary part of the cloaca, which becomes separated from the hindgut by downward growth of the urorectal septum, forms the largest part;
- the pelvic end of the mesonephric ducts and their ureteric buds form the trigone;
- the allantois, a midline diverticulum which grows from the cloaca into the connecting stalk, forms the apex of the bladder.

Ascent of the kidneys occurs before the various parts of the gut and its mesentery become attached to the posterior abdominal wall. The kidneys and ureters therefore lie behind the bowel and the mesenteric vessels that supply the bowel.

Occasionally, failure of ascent occurs and a pelvic kidney is formed; this is often because the right and left kidney primordia fuse caudally to form a 'horseshoe' kidney. Other evidence of the ascent of the kidneys may be found in variations of the vasculature of the adult organ (see below).

In the adult, remnants of the allantois form a fibrous cord (urachus, median umbilical ligament) between the apex of the bladder and the umbilicus.

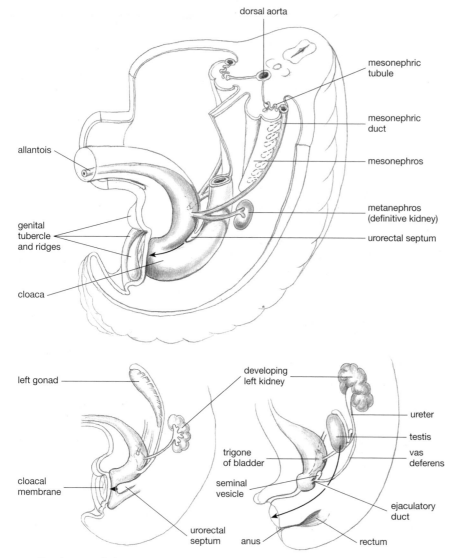

dorsal aorta

mesonephric tubule

mesonephric duct

mesonephros

allantois

metanephros (definitive kidney)

urorectal septum

genital tubercle and ridges

cloaca

left gonad

developing left kidney

ureter

testis

vas deferens

trigone of bladder

seminal vesicle

ejaculatory duct

cloacal membrane

urorectal septum   anus

rectum

6.7.1 Development of urinary system and hindgut.

Qu. 7A *If a human fetus developed without both kidneys, could metabolic wastes be excreted?*

Kidneys

The kidneys produce urine by ultrafiltration of the blood plasma and subsequent modification of the ultrafiltrate by reabsorption and secretion. Like the liver and spleen, the kidneys are highly vascular and firm but pliable in consistency. They lie on the posterior abdominal wall in the paravertebral gutters, the right a little lower than the left because of the mass of the liver. They are oriented obliquely, with their medial borders anterior to the lateral borders and their upper poles nearer the midline than the lower poles (**6.7.2**).

Qu. 7B *How does the obliquity of the kidneys alter their apparent size on antero-posterior abdominal radiographs?*

The **hilum** of each kidney, a prominent feature of the medial border, lies at the level of the transpyloric plane, about 5 cm lateral to the midline. It transmits, from anterior to posterior:

- the renal vein;
- the renal artery, with accompanying autonomic nerves and lymphatics;
- the pelvis of the ureter.

The kidneys lie retroperitoneally, behind the parietal peritoneum. Each has a thin **capsule** around which is a variable amount of encapsulated **perinephric fat**, which in the obese also surrounds the adrenal gland adjacent to the upper pole of each kidney.

The upper poles of the kidneys lie at the level of the twelfth rib (and eleventh rib on the left), on muscular fibres of the diaphragm. They therefore overlap the lower extremities of the pleura and pleural spaces.

The lower parts of the kidneys lie, from medial to lateral, on psoas, quadratus lumborum, and transversus abdominis. Immediately below the twelfth rib is the subcostal nerve (T12) and, below that, the anterior primary ramus of L1, which divides into iliohypogastric and ilio-inguinal nerves. These nerves all lie in direct contact with the perinephric fat before piercing transversus abdominis to reach the neuromuscular plane.

Qu. 7C *What might be the consequences of a stab wound in the back between the left eleventh and twelfth ribs?*

The anterior relations of the two kidneys differ (**6.7.2**). The upper part of the *right* kidney, with the adrenal gland on its upper pole, lies behind the right lobe of the liver (partly in its bare area, and partly separated by the hepatorenal pouch), the second part of the duodenum curving around the head of the pancreas lies on the hilum of the right kidney; the hepatic flexure of the colon contacts its lower pole.

The *left* kidney, with its adrenal gland, lies in the floor of the lesser sac, behind the stomach. It is crossed by the body of the pancreas. The left kidney also contacts the spleen, the splenic flexure of the colon, and, at its lower pole, coils of jejunum.

In the fetus and infant, the pelvic cavity is very small and the bladder, which is almost tubular in shape, lies almost entirely above the pelvic brim, in contact with the anterior abdominal wall. The pelvis deepens as the child grows, so that, by the age of about 6 years, the bladder lies entirely in the pelvis.

The development of the urinary tract and genital systems is closely linked. In males, the urinary tract retains its connection with the mesonephric duct, which forms the vas deferens, and both the genital and urinary systems use the male urethra; the development of the penis and penile urethra (p. 205) transfers the external urethral orifice from the perineum to the tip of the penis. In females, the mesonephric duct degenerates and the urinary and reproductive systems therefore become separate.

Developmental anomalies can result in the formation of abnormal connections (fistulae) between the urinary system and other pelvic organs (rectum, vagina), and of abnormal orifices of the urinary tract, particularly in the male urethra.

Examination of a kidney, sliced vertically through the hilum from medial to lateral (6.7.3), shows an outer pale **cortex** and an inner **medulla** with dark triangular masses (pyramids), the apices (papillae) of which appear striated. This appearance is due to the presence of numerous collecting tubules which carry urine to the funnel-shaped **minor calyces** into which the papillae project. The 7–13 minor calyces join to form 2–3 **major calyces** which lead into the **pelvis** of the ureter.

The renal pelvis, ureter, and bladder are lined with specialized urine-resistant **urothelium** (transitional epithelium). The pelvis has two muscle coats: the outer surrounds the pelvis and calyces; the inner also surrounds the calyces but ends at the junction between the major calyces and the pelvis. This inner layer is thought to act as a **pacemaker** for the initiation of the peristaltic waves which spread downward, conveying drops of urine through the ureter to the bladder.

Blood supply, lymphatic drainage, and innervation of the kidneys

Two **renal arteries** (6.7.2, 6.7.3b, 6.7.4) originate from the sides of the aorta, close to the midline superior mesenteric artery. They are short and wide, the right passing behind the inferior vena cava to reach the right kidney. Before entering the hilum of the kidney each artery divides into about five branches (lobar arteries) which are effectively end arteries. The lobar arteries pass through the medulla then give rise to arcuate arteries which run along the cortico-medullary junction. The arcuate vessels give rise to arteries which radiate outward toward the kidney capsule giving off afferent arterioles to the glomeruli.

The afferent arterioles form a tuft of capillaries (glomerulus) which invaginates the Bowman's capsule of a nephron to form a renal corpuscle. The glomerular capillaries from superficial renal corpuscles do not drain into a vein, but gather together to form an efferent arteriole which leaves the glomerulus and supplies the capillaries which surround the convoluted tubules in the cortex. This capillary plexus then drains into veins.

The efferent arterioles from renal corpuscles which lie close to the medulla (juxtamedullary nephrons) do not join the cortical capillary plexus, but each gives rise to a leash of straight vessels (vasa recta) that pass deeply into the medulla around the long tubular loops of Henle which arise from these nephrons. The veins pass back up to join arcuate veins at the cortico-medullary junction. This arrangement ensures the slow flow of blood in capillaries of the medulla that is essential for the development and maintenance of the medullary osmotic gradient.

An appreciation of the microscopic structure of the tubules and vasculature of the kidney is essential to an understanding of the mechanism of production of urine.

An 'accessory' renal artery may enter the lower (or less commonly upper) pole of a kidney. However, this

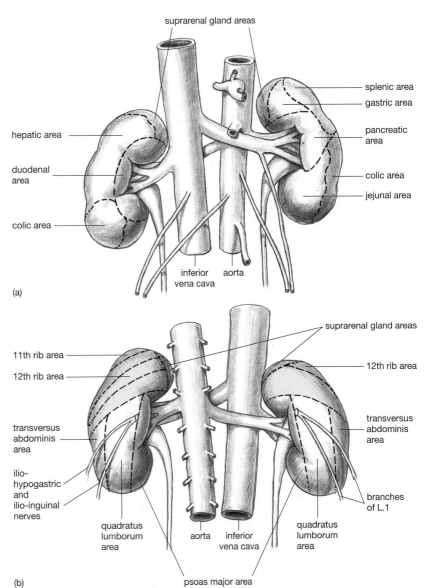

6.7.2 (a) Anterior and (b) posterior relations of the kidneys.

name is misleading because, like branches of the main renal artery, such vessels are end arteries which supply one lobe of the kidney. An accessory renal artery to the lower pole often passes close to the pelvi-ureteric junction and could obstruct the flow of urine.

Qu. 7D *What does the presence of these vessels reflect?*

The large **renal veins** drain into the inferior vena cava. The right passes directly into the inferior vena cava; the left is longer and usually runs over the anterior surface of the aorta. The left adrenal vein and left gonadal vein drain into the left renal vein; veins from the right adrenal and right gonad drain directly into the inferior vena cava. This reflects the presence, during development, of venous channels on either side of the aorta, with interconnecting anastomoses, and the subsequent suppression of the channel on the left.

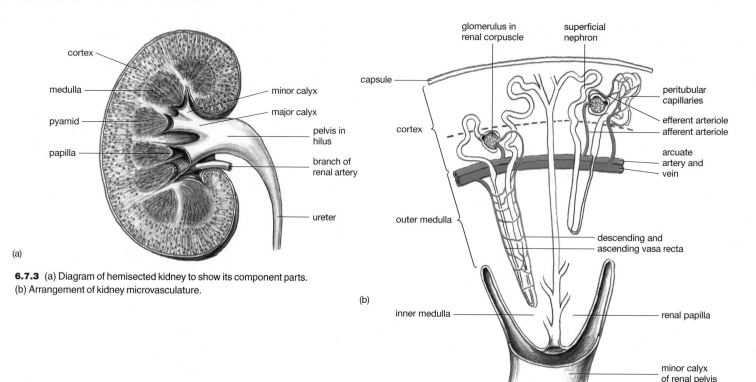

(a)

6.7.3 (a) Diagram of hemisected kidney to show its component parts.
(b) Arrangement of kidney microvasculature.

(b)

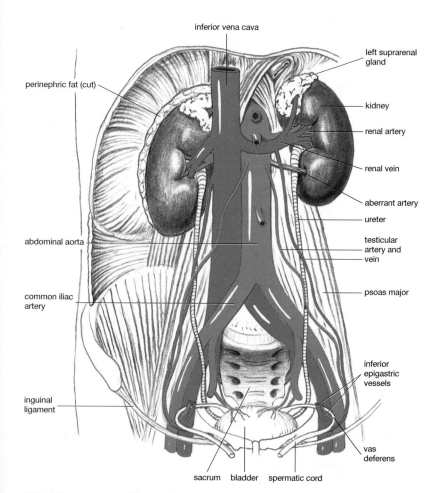

6.7.4 Urinary system and its vessels.

Lymph from the kidneys drains via nodes in the kidney hilum to para-aortic nodes around the renal arteries. These drain through right and left lumbar lymphatic trunks into the cisterna chyli.

The kidneys receive **sympathetic** fibres from the renal plexus; these regulate the renal vessels and provide one stimulus for the secretion of renin from the juxtaglomerular apparatus. The role of the vagal fibres to the kidney is uncertain.

Ureters

The ureters are narrow tubes (about 25 cm long) with thick muscular walls which convey urine from the kidney to the bladder. At operation they can be identified easily by the peristaltic waves that occur from time to time and can be induced by gently pinching the ureter with a pair of forceps.

Kidney transplantation

Transplantation of a kidney is now a definitive treatment for kidney failure. Kidneys are removed from compatible living relatives or from donors who die from trauma or haemorrhage that irreversibly damages the brain.

The new kidney is usually placed in the iliac fossa, where its artery and vein can conveniently be anastomosed to the external iliac vessels. The ureter is inserted obliquely through the bladder wall (see below).

Each ureter passes downward on psoas from the renal pelvis to the pelvic brim, covered by parietal peritoneum. On both sides, the ureters are crossed by the gonadal vessels; the left ureter is also crossed at the pelvic brim by the superior rectal artery in the apex of the sigmoid mesocolon. At the sacroiliac joint each ureter crosses the bifurcation of the common iliac artery to pass on to the side wall of the pelvis.

At the level of the ischial spine, each ureter runs forward and medially in the fascia of the pelvic floor to reach the lateral angle of the bladder. In the male the ureter, just before it enters the bladder, is crossed by the vas deferens, which passes downward behind the bladder; in the female the ureter is crossed by the uterine artery in the base of the broad ligament.

Each ureter receives an anastomosing arterial supply from:
- the renal artery to its upper end;
- the gonadal artery;
- the common iliac artery at the pelvic brim;
- the superior vesical artery to its lower end.

The ureters receive autonomic fibres of both types via the renal, aortic, presacral, and pelvic plexuses (see 2.13, 6.3.10).

Urinary bladder

The bladder is a distensible reservoir for the urine, which is produced continuously by the kidneys and conveyed to the bladder by peristaltic contractions of the ureters. The bladder is emptied into the urethra periodically by contraction of its muscular wall. It lies in the most anterior part of the pelvis and, when empty of urine, has the form of an inverted triangular pyramid (6.7.5).

- Its superior surface—**fundus**—extends from the **apex**, which lies behind the symphysis pubis, to the ureters, which enter the bladder at its postero-lateral angles. From the apex a fibrous cord (the median umbilical ligament, derived from the urachus) extends in the midline up to the umbilicus.
- Its posterior surface—**base**—extends from the two ureters down to the internal urethral orifice. Internally, it is formed largely by the trigone.
- Its two triangular side walls meet in the midline from the apex to the internal urethral orifice. They lie in contact with the pelvic walls formed largely by levator ani. In the midline, anteriorly, the bladder lies in contact with the pubic symphysis, separated by a potential **retropubic space**.

The ureters pass obliquely through the bladder wall at each upper, outer extremity of the base.

Qu. 7E *What is the function of this oblique course?*

Bladder filling

As the bladder fills, its base changes little in size or shape, but the fundus expands upward. Normally it is emptied before its upper margin rises above the pelvic brim. The fundus is covered by peritoneum which is reflected from the anterior abdominal wall. This peritoneum is only loosely attached to the anterior abdominal wall so that, if the bladder becomes abnormally full, it expands upward between the peritoneum and the abdominal wall.

Suprapubic access to the bladder

If the bladder cannot be emptied via the urethra, for example because of a grossly enlarged prostate gland (see below), it can expand very considerably, stripping the peritoneum from the abdominal wall as far up as the umbilicus in the most extreme cases. If it cannot be drained by passing a catheter along the urethra, it therefore can be drained, without entering the peritoneal cavity, by inserting a catheter through the lower part of the anterior abdominal wall and into the anterior surface of the distended bladder (suprapubic cystotomy). Similarly, an enlarged prostate which has to be removed can be approached surgically via a suprapubic approach without opening the peritoneal cavity. However, more commonly operations on enlarged prostates are now done by transurethral endoscopic surgery.

In males, the peritoneum is reflected from the superior surface of the bladder posteriorly on to the middle part of the rectum (6.7.6), thereby forming a **rectovesical pouch**. In females, the uterus lies behind the bladder and is normally bent forward (anteflexed) over its superior surface. Peritoneum therefore covers only a part of the upper surface of the bladder, but is reflected on to the uterus, forming a shallow uterovesical pouch; it then covers the uterus and posterior fornix of the vagina before being reflected on to the middle part of the rectum to form a deep **recto-uterine pouch** (pouch of Douglas) (6.9.3; p. 193). Coils of small bowel and/or sigmoid colon lie above the bladder and occupy these pouches.

In the female, the base of the bladder is related posteriorly to the cervix of the uterus and anterior wall of the vagina, and the bladder neck lies on the pelvic floor. In the male, the bladder neck is separated from the pelvic floor by the prostate gland (p. 184). On each side, the vas deferens crosses over the ureter to reach the base of the bladder where it lies medial to the seminal vesicle; it joins with the duct of the seminal vesicle then penetrates the prostate to enter into the urethra. The seminal vesicle and vas deferens lie between the bladder base and rectum enclosed in a sheet of fascia (rectovesical fascia; a remnant of the urorectal septum) which provides a plane used by surgeons to separate the prostate from the rectum.

In both sexes thickenings of the pelvic floor fascia pass from the pubic symphysis to the region of the bladder neck. In the male these are called **puboprostatic ligaments**; in the female, **pubovesical ligaments**. They support the neck of the bladder and play a role in the maintenance of urinary continence (p. 173).

Transplantation of the ureters

If the bladder becomes blocked by carcinoma, it may be necessary to transplant the distal end of the ureter from the bladder. The ureters may be exteriorized on to the abdominal skin via a conduit of ileum. Great care must be taken to protect the skin surrounding the stoma, because skin is not resistant to damage by urine.

Ureteric colic

Pain originating from spasm of the ureters (ureteric colic) caused by the passage of urinary stones is very severe and is referred to cutaneous areas on the same side of the abdomen and not to the midline. It usually starts in the loin (T10) and passes round to the groin, ending in the scrotum or labia (L1).

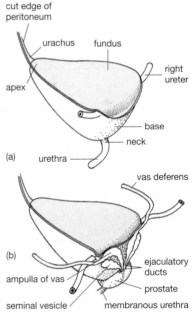

6.7.5 Diagrams of bladder showing its relations to peritoneum in (a) female (b) male.

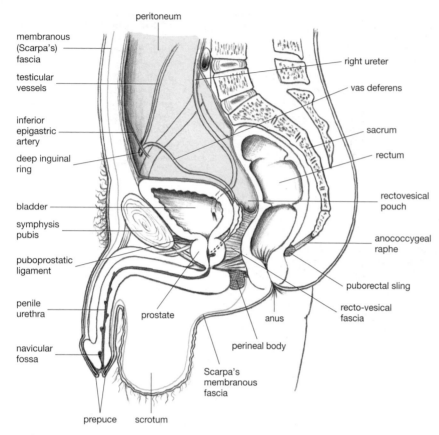

peritoneum

membranous
(Scarpa's)
fascia

testicular
vessels

inferior
epigastric
artery

deep inguinal
ring

bladder

symphysis
pubis

puboprostatic
ligament

penile
urethra

navicular
fossa

prepuce   scrotum

prostate

anus

perineal body

Scarpa's
membranous
fascia

right ureter

vas deferens

sacrum

rectum

rectovesical
pouch

anococcygeal
raphe

puborectal sling

recto-vesical
fascia

6.7.6 Sagittal section of male pelvis.
Note: the rectum contains faeces only intermittently before defaecation.

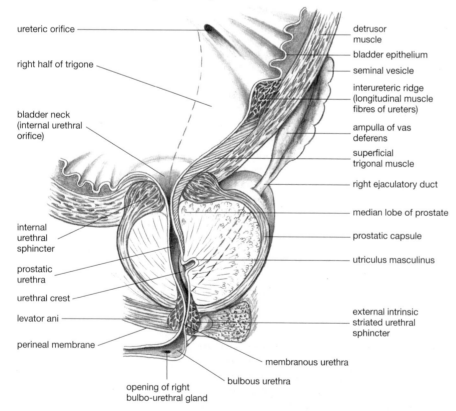

ureteric orifice

right half of trigone

bladder neck
(internal urethral
orifice)

internal
urethral
sphincter

prostatic
urethra

urethral crest

levator ani

perineal membrane

opening of right
bulbo-urethral gland

bulbous urethra

membranous urethra

detrusor
muscle

bladder epithelium

seminal vesicle

interureteric ridge
(longitudinal muscle
fibres of ureters)

ampulla of vas
deferens

superficial
trigonal muscle

right ejaculatory duct

median lobe of prostate

prostatic capsule

utriculus masculinus

external intrinsic
striated urethral
sphincter

6.7.7 Detail of sagittal section of male bladder and prostate to show the prostate gland, urethra, and internal and external urinary sphincters.

Internal features of the bladder

Inside the bladder, the **ureteric orifices** appear as small slits at either end of a muscular interureteric bar. This marks the upper limit of the smooth **trigone**, which extends down the posterior wall of the bladder to the **bladder neck** and **internal urethral orifice**. The mucosa lining the fundus and side walls of the empty bladder is thrown into a series of folds which allow for expansion. By contrast, the mucosa covering the trigone is smooth and firmly attached to the fibrous base plate because this part of the bladder does not expand.

Bladder emptying (micturition)

The smooth muscle wall of the bladder, **detrusor** (6.7.7), is arranged in three rather indistinct layers (outer and inner longitudinal; middle circular) of intermeshed bundles of non-striated muscle cells. Its contraction expels urine through the internal urethral orifice into the urethra. Obstruction of the outflow results in hypertrophy and 'trabeculation' of the bladder musculature. Contraction of the detrusor not only increases intravesical pressure but also alters the geometry of the bladder neck and trigone, creating a funnel shape which opens the bladder neck. This can be visualized by taking radiographs of a subject voiding urine which contains radiopaque material (a 'micturating cystogram'; see 6.7.8). Note the marked change in the fundus and relative lack of change in the base as the bladder empties.

The micturition reflex involves sensory nerves from the bladder which pass to the sacral spinal cord and parasympathetic innervation to the detrusor muscle (p. 231).

Continence of the bladder neck (internal urethral orifice) is normally maintained by a functional 'internal urethral sphincter'. The exact nature of this is still debated, and it differs in males and females. In males there is a distinct circular collar of smooth muscle above the prostate, but in females the smooth muscle fibres do not form a distinct circular sphincter. One suggestion is that the arrangement of the three layers of muscle in relation to the trigone keeps the trigone pressed forward against the anterior wall of the bladder neck (6.7.7).

Qu. 7F *What other abnormality might a micturating cystogram show that would not be seen if micturition was not in progress?*

Blood supply and innervation of the bladder

The bladder receives arterial blood from the internal iliac artery through the **superior vesical artery**, which supplies its fundus, and the **inferior vesical artery**, which supplies its base. Its venous drainage passes via corresponding vessels to the inferior vena cava.

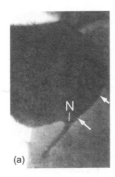
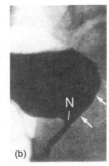
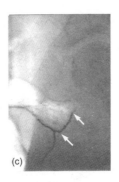

(a)   (b)   (c)

6.7.8 Micturating cystogram. Note funnel shape of neck of bladder (N) in (a), (b); also the constant position of the base of the bladder (*arrows*).

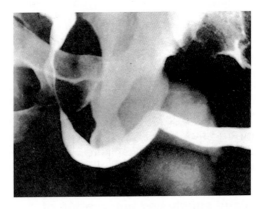

6.7.9 Male urethra filled with contrast material.

The bladder receives autonomic nerves via the pelvic plexuses on either side of the rectum. Stimulation of the efferent **parasympathetic** fibres from the pelvic splanchnic nerves (see **6.8.6, 8.5**) results in contraction of the detrusor and emptying of the bladder, but the role of sympathetic fibres in micturition is uncertain (they may delay micturition by inhibiting the detrusor). Afferents conveying the sensation of **distension** seem to run with sympathetic pathways to upper lumbar segments of the spinal cord, whereas those conveying vesical **pain** pass with sympathetic and parasympathetic fibres, respectively to upper lumbar and sacral segments.

Urethra

In the *female*, the urethra is a relatively short, straight muscular walled tube which passes downward and a little forward from the bladder neck to the external urethral orifice in the vestibule (p. 209). Except during the passage of urine, its walls are in contact. Just before it passes through the perineal membrane it is surrounded by the voluntary **external urethral sphincter** (see below). The urethra is firmly attached to the anterior wall of the vagina and its external orifice forms an antero-posterior slit just in front of the vagina. The short length of the female urethra may be responsible for the greater incidence of bladder infection (cystitis) in women compared to men.

Qu. 7G *What functional consequences might prolapse of the anterior vaginal wall have for the urinary system?*

In the *male*, the urethra is much longer, and extends from the bladder neck to the tip of the penis, passing through the prostate and the entire corpus spongiosum of the penis. As in the female, its lumen is closed except during the passage of urine and semen. It has three parts:

- The **prostatic urethra** runs downward through the prostate. This is the widest and most dilatable part and its lumen is made crescentic in transverse section by the midline posterior **urethral crest** with lateral prostatic sinuses into which the tiny ducts of the prostate open (see **6.7.15b**). The **seminal colliculus** is an elevation of the middle of the urethral crest through which the ejaculatory ducts pass on either side of the midline to join the urethra, thus bringing together genital and urinary systems. A tiny blind-ending **prostatic utricle** also opens on to the seminal colliculus; it is a remnant of the lower end of the paramesonephric ducts (p. 181).

- The **membranous urethra** (**6.7.7, 6.7.9**) starts at the lower end of the prostate and passes first through the pelvic floor and then through the perineal membrane. In both sexes, the voluntary striated muscle of the **external urethral sphincter**, which is innervated by the pudendal nerve, lies between these two sheets. The membranous urethra is the shortest and least dilatable part of the male urethra. Two bulbo-urethral glands lie on either side of the membranous urethra.

- The **penile urethra** (**6.7.9**) starts immediately below the perineal membrane. It is enclosed in the corpus spongiosum of the penis. In the perineum its proximal part is dilated to form the **bulbous urethra**. The ducts of the bulbo-urethral glands pierce the perineal membrane and open into the bulbous urethra. The remainder of the penile urethra is a narrow transversely oriented slit, which expands within the glans penis to form the **navicular fossa**, a vertical slit which opens as the external urethral orifice on to the tip of the glans. This variation in the diameter and orientation of the urethra ensures a steady stream of urine during micturition. Mucous **urethral glands** open into the penile urethra via ducts, the mouths of which are directed towards the external urethral orifice.

Qu. 7H *At what part of the male urethra is it most likely that difficulty in passing a catheter will be encountered?*

Blood supply and innervation of the urethra

The arterial supply of the urethra is from perineal branches of the **internal pudendal artery** (p. 210); its venous drainage is along corresponding veins to the internal iliac vein and thence to the inferior vena cava.

Lymphatic drainage of the urethra passes with that of the penis (p. 210).

The prostatic urethra is supplied with autonomic nerves from the pelvic plexuses but the penile urethra receives somatic sensory innervation from perineal branches of the pudendal nerve (S2, 3, 4).

Adrenal (suprarenal) glands

The adrenal glands are paired endocrine organs situated on the upper pole of each kidney. Each comprises an outer cortex which secretes steroids and an inner medulla which secretes catecholamines. The right adrenal is triangular in shape and lies between the diaphragm and the bare area of the liver, just lateral to the inferior vena cava. The left is more crescentic in shape and is placed more medially over the upper pole of the left kidney; it lies on the left crus of the diaphragm and coeliac plexus behind the lesser sac and stomach. Both receive a profuse arterial supply from adjacent vessels, and sympathetic innervation from thoracic splanchnic nerves, which consist mainly of preganglionic fibres to the adrenal medulla. See Chapter 7.

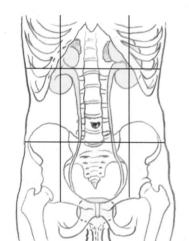

(a)

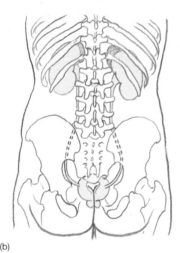

(b)

6.7.10 Surface markings of urinary system and suprarenal glands (a) anterior (b) posterior. The bladder is shown distended.

Living anatomy

With your subject lying supine on an examination couch, draw on the anterior abdominal wall the surface projection (**6.7.10**) of the kidneys (not forgetting the foreshortening effect of their oblique orientation), the ureters, and the bladder. Indicate, too the position of the right and left adrenal glands.

Stand to the right side of the examination couch and try to feel the lower pole of your subject's left kidney. To do this, place the palm of your left hand under the left posterior abdominal wall, lifting it and the kidney forward to meet your examining right hand. Repeat the examination on the right kidney. Only in very lean subjects is it usually possible to feel the kidneys distinctly.

Next, with your colleague lying prone, mark the position of the twelfth and the eleventh ribs and draw the surface projection of the kidneys (**6.7.10b**, **6.7.2**). Add the markings of the lower limits of the pleura and lungs.

Imaging the urinary tract

Radiology

Examine a plain antero-posterior radiograph of the abdomen (see **6.1.10**). It may be possible to make out the outline of the kidneys lying lateral to and parallel with the lateral borders of psoas major, and especially their lower poles, but the remainder of the urinary system cannot be visualized unless radiopaque material has been introduced into its lumen. This can be done by administering intravenously a radiopaque substance which is excreted by the kidneys (an intravenous ureterogram, or IVU; **6.7.11**).

Alternatively, radiopaque material can be introduced into the urinary system retrogradely, either into the bladder, or through ureteric orifices and into the ureters (retrograde cystogram or ureterogram; **6.7.12**).

Intravenous urography shows whether or not each kidney is able to excrete urine, in addition to defining the lower urinary tract. Examine the urograms, noting the times after injection of contrast material when the radiographs were taken. In **6.7.11a** contrast material is present in the renal parenchyma; in **6.7.11b** it outlines the minor calyces with the medullary pyramids projecting into them, the pelvis and the ureter passing downward from it to the bladder; **6.7.11c** shows the form of the full bladder.

Retrograde ureterograms (**6.7.12**) show the ureter and pelvis even if the kidney cannot excrete the contrast medium.

Qu. 7I *What bony landmarks does the ureter follow as it passes to the bladder?*

Examine the radiograph (**6.7.9**) of the male urethra filled with contrast material. Such radiographs are obtained by filling the bladder with contrast material, then getting the subject to attempt to micturate while preventing efflux of urine by clamping the penis.

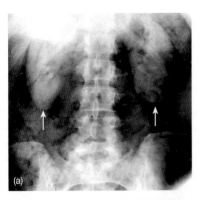

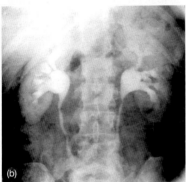

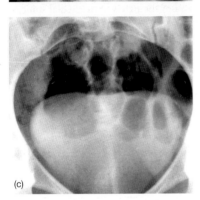

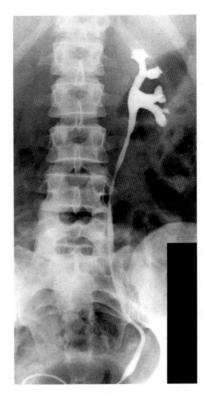

6.7.11 Intravenous urogram (a), (b), (c). In (a) note the lower pole of right and left kidneys (arrows).

CT, MRI, and ultrasound

The kidneys, ureters, and bladder can be visualized by ultrasound (see **3.12**) and by computed tomography (**6.13.1**).

Endoscopy

The interior of the urethra and bladder can usually be examined by a rigid or a flexible endoscope passed through the urethra. The photographs show the interior of the penile urethra (**6.7.15a**), the prostatic urethra (**6.7.15b**) and urethral crest, the opening of a ureter into the bladder (**6.7.15c**) and the fronds of a bladder tumour (**6.7.15d**).

Questions and answers

Qu. 7A *If a human fetus developed without kidneys, could metabolic wastes be excreted?*

6.7.12 Retrograde ureterogram.

Urinary stones

Calculi (stones) may form in the urinary system (as in the biliary system). They can be found in the renal pelvis (**6.7.13**), in the ureter (where they cause great pain), or in the bladder (**6.7.14**). Calculi tend to become impacted at narrow points, viz. the pelvi-ureteric junction, the part of the ureter that crosses the common iliac artery, and the entrance to the bladder. Calcified calculi are visible on a plain abdominal radiograph; others are radiolucent. It is clearly necessary to be able to trace the course of the ureter on radiographs in order to decide whether an opacity visible on a plain radiograph could be a calculus within the urinary system. The introduction of contrast material is also used to define calculi, particularly those that are radiolucent.

Answer A fetus without kidneys can excrete waste via the placenta, but most die *in utero*. The volume of amniotic fluid may be small (oligohydramnios) because fetal urine contributes to the amniotic fluid. If only one kidney forms (unilateral agenesis) it hypertrophies and normal living is possible. A number of genes, including *WT1* (Wilms tumour 1) are now known to be necessary for the development of the kidneys. If *WT1* is inactivated, kidney tumours develop in childhood.

Qu. 7B *How does the obliquity of the kidneys alter their apparent size on antero-posterior abdominal radiographs?*

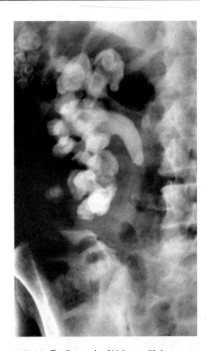

6.7.13 Radiograph of kidney with large calculus filling a distended renal pelvis.

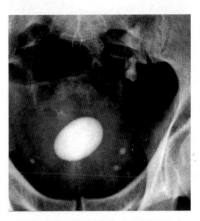

6.7.14 Radiograph to show a bladder stone.

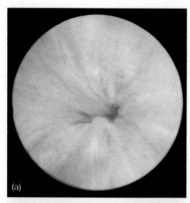

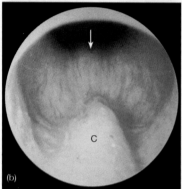

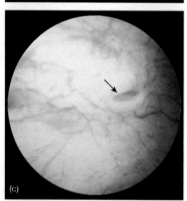

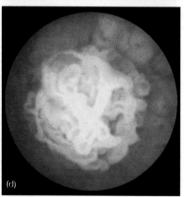

6.7.15 Endoscopic views of: (a) penile urethra; (b) prostatic urethra showing urethral crest (C) and bladder neck (*arrowhead*); (c) bladder wall, showing ureteric orifice (*arrow*) (high-power view); (d) bladder tumour with numerous fronds, arising from the smooth normal bladder wall.

Answer The oblique orientation of the kidneys (their medial borders lie anterior to their lateral borders) will, on an A-P radiograph of the abdomen, give the impression that their transverse diameter is narrower than it really is.

Qu. 7C *What might be the consequences of a stab wound in the back between the left eleventh and twelfth ribs?*

Answer A stab wound between the eleventh and twelfth ribs on the left side might pierce the costodiaphragmatic recess of the pleura, the diaphragm, the parietal peritoneum, the upper pole of the left kidney, and possibly the spleen.

Qu. 7D *What does the presence of these vessels reflect?*

Answer The separate lower pole arteries reflect persistence of lower vessels formed during ascent of the kidney during development.

Qu. 7E *What is the function of this oblique course?*

Answer When the bladder fills, the oblique entry of the ureters into the muscular wall of the bladder acts as a valve to prevent reflux of urine back up the ureters. Similarly, during micturition, contraction of bladder wall musculature compresses the oblique ureterovesical junction and prevents reflux.

Qu. 7F *What other abnormality might a micturating cystogram show that would not be seen if micturition was not in progress?*

Answer A cystogram taken during micturition may show reflux of urine into the ureter(s) if the ureterovesical junction is incompetent.

Qu. 7G *What functional consequences might prolapse of the anterior vaginal wall have for the urinary system?*

Answer Because the ureter is firmly attached to the anterior wall of the vagina, any distortion of the vagina also affects the urethra. This frequently causes urinary incontinence, most often when intra-abdominal pressure is raised (stress incontinence).

Qu. 7H *At what part of the male urethra is it most likely that difficulty in passing a catheter will be encountered?*

Answer In a young healthy male there is usually no difficulty in passing a catheter into the bladder, although a little resistance might be encountered at the junction between the bulbous and membranous urethra.

Qu. 7I *What bony landmarks does the ureter follow as it passes to the bladder?*

Answer The ureter descends on psoas close to the tips of the lumbar transverse processes; it crosses the pelvic brim close to the sacroiliac joint then passes across the lateral wall of the pelvis to the ischial spine, where it turns medially to reach the upper lateral angle of the bladder.

Male reproductive system

Male reproductive system

The male reproductive system consists of the testes, which produce the male gametes (spermatozoa) and male sex hormones, the epididymis and vas deferens, which convey the spermatozoa centrally, and the seminal vesicles and prostate, which secrete the seminal fluid in which the spermatozoa are suspended. The resultant semen is ejected through the urethra during emission and ejaculation. The male urethra is extended and surrounded by erectile tissue to form the penis, which enables semen to be deposited as close as possible to the uterus during sexual intercourse. The male urethra and penis are therefore not only a part of the urinary system but also essential parts of the male reproductive system.

Development

Sexual differentiation

By the fourth week of fetal life primordial germ cells which will eventually give rise to the gametes can be distinguished close to the root of the yolk sac. They migrate on to the posterior abdominal wall, multiplying by mitosis *en route*, until they reach the gonadal ridge component of the intermediate mesoderm on either side, where the potential gonads form medial to the mesonephros. Certain genes are essential for the differentiation of the intermediate mesoderm to form a gonadal primordium. These are the Wilms tumour gene (involved in some kidney tumours) and the steroidogenic factor 1 gene (needed for steroid-producing tissues). By contrast, the presence of the primordial germ cells is not essential, but the resultant 'gonads' could not produce gametes.

Embryos with XY sex chromosomes normally possess on the Y chromosome a 'testis-determining gene' (*SRY*) the expression of which causes the mesoderm of the gonadal primordium to differentiate into Sertoli cells, and the gonad to develop as a testis. The Sertoli cells and primordial germ cells become enclosed by basement membrane to form primitive seminiferous tubules. The Sertoli cells secrete a glycoprotein hormone (anti-Müllerian hormone, AMH; or Müllerian inhibitory substance, MIS). This acts locally to suppress development of the paramesonephric (Müllerian) system which would otherwise form the female reproductive tract. In the interstitial mesoderm between the tubules, clusters of Leydig cells differentiate and, by the seventh week, begin to secrete testosterone. This androgenic steroid diffuses locally to stimulate differentiation and growth of the male reproductive tract from the mesonephric (Wolffian) system. Circulating testosterone stimulates growth and differentiation of the prostate and seminal vesicles (see below) and is needed for the differentiation of male external genitalia. This occurs rather late in development (starting about the eighth week of fetal life) and requires conversion of circulating testosterone by target tissues to the more potent androgen 5α-dihydrotestosterone. During fetal life differential growth carries the testes down the posterior abdominal wall to the inguinal region and then, around birth and stimulated by testosterone, the testes descend into the scrotum. Some parts of the brain, particularly some hypothalamic nuclei, also become sexually dimorphic. Interestingly, some of these effects require the conversion of testosterone to oestrogen by the brain.

If the embryo has an XX genotype, the *SRY* gene is normally absent and the mesodermal gonadal cells differentiate to become granulosa cells. These surround the primordial germ cells to form primordial follicles. Even in the absence of an ovary and its hormones, subsequent development of the reproductive tract and external genitalia is female. For this reason the female has been called the 'default' pattern in mammals, but it, too, requires expression of an appropriate set of genes. In normal females the ovarian oestrogenic steroids and their receptors are important primarily at, and after, puberty.

Development of the male reproductive tract

In the male, although most of the mesonephros degenerates, some mesonephric tubules lying close to the differentiating gonad persist. These become the efferent ducts which connect the seminiferous tubules with the mesonephric duct (**6.8.1**). This duct grows under the influence of testosterone to form the epididymis

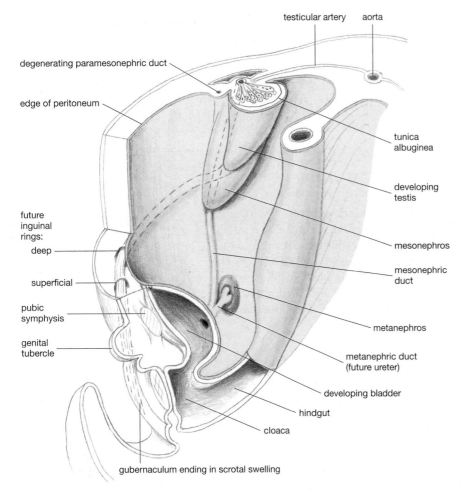

testicular artery   aorta

degenerating paramesonephric duct

edge of peritoneum

tunica albuginea

developing testis

future inguinal rings:

deep

mesonephros

mesonephric duct

superficial

pubic symphysis

genital tubercle

metanephros

metanephric duct (future ureter)

developing bladder

hindgut

cloaca

gubernaculum ending in scrotal swelling

6.8.1 Development of male urogenital system.

and vas deferens. At the distal end of each mesonephric duct a ureteric bud (metanephric duct), which induces formation of the metanephric kidney, appears.

The rotation of the caudal ends of the mesonephric and metanephric ducts as they are incorporated into the wall of the bladder (p. 169) means that the mesonephric ducts enter the part of the urogenital sinus that will narrow to form the proximal (prostatic) part of the urethra.

Stimulated by androgens, the prostate develops as numerous glandular outgrowths from the urethra, and each seminal vesicle develops as a lateral outgrowth from the terminal part of the vas deferens (mesonephric duct). The seminal vesicle and vas deferens on each side therefore open into a common ejaculatory duct which passes obliquely through the lateral aspects of the developing prostate to open into the prostatic urethra.

Descent of the testis

As a result of differential upward growth of the fetal trunk, the testis, which began its development high on the posterior abdominal wall, comes to lie close to the inguinal region.

A band of specialized tissue, the **gubernaculum**, which is attached to the lower pole of the testis and to scrotal skin, acts as a 'pathfinder' for testicular descent. As the testis, together with its vessels and nerves, descends through the fascia and muscles of the inguinal region of the abdominal wall, it is preceded by a diverticulum of the peritoneal cavity, the **processus vaginalis** (6.2.1). The distal end of this diverticulum remains patent to form the bursa-like **tunica vaginalis**, which almost completely envelops the testis; its proximal end in the inguinal canal normally closes anterior to the spermatic cord.

The scrotum

The **scrotum** (6.8.2) is a pouch of skin which contains, on each side, the descended testis and epididymis at a temperature compatible with the formation of spermatozoa (spermatogenesis).

Scrotal subcutaneous tissue has very little fat and is made reddish in colour by the presence of smooth muscle fibres (**dartos**) which wrinkle the scrotal skin to bring the testes closer to the warmth of the trunk.

Beneath dartos is a fascial sac formed by the continuation of the membranous layer of superficial fascia of the lower abdominal wall. This sac is sealed laterally and posteriorly by attachments to the ischial rami, the perineal body, and the posterior margin of the perineal

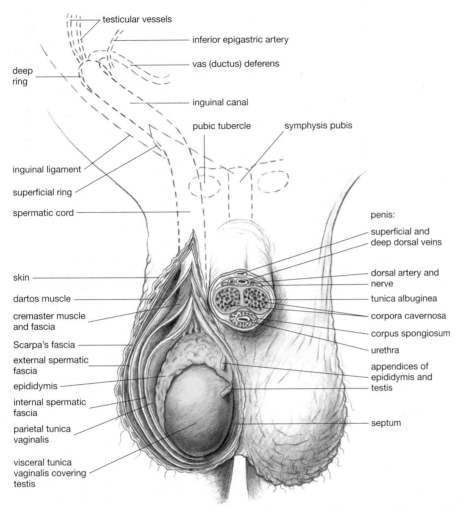

testicular vessels

inferior epigastric artery

vas (ductus) deferens

deep ring

inguinal canal

pubic tubercle   symphysis pubis

inguinal ligament

superficial ring

spermatic cord

penis:

superficial and deep dorsal veins

skin

dorsal artery and nerve

dartos muscle

tunica albuginea

cremaster muscle and fascia

corpora cavernosa

Scarpa's fascia

corpus spongiosum

external spermatic fascia

urethra

epididymis

appendices of epididymis and testis

internal spermatic fascia

parietal tunica vaginalis

septum

visceral tunica vaginalis covering testis

6.8.2 Coverings of testis, spermatic cord, and penis. The layers covering the cord and scrotum are artificially separated; they are difficult to distinguish by dissection.

membrane (p. 206). A fibrous raphe divides the scrotum into right and left compartments.

Qu. 8A *Of what perineal pouch is this sac a part (see p. 208)?*

Within each side of this fascial sac the testis, epididymis, vessels, and tunica vaginalis are surrounded by a series of fused layers derived from the three muscle layers of the abdominal wall (**6.8.2**; see **6.2.2**). That from internal oblique forms loops of **cremaster** muscle which spiral around the cord and can raise and lower the testis to control its temperature. The cremasteric reflex (see **8.3**) can be elicited by touching the medial aspect of the thigh.

Within these fibromuscular coverings, the parietal and visceral layers of the **vaginal tunic** (tunica vaginalis) enclose a bursa-like space covering the anterior and lateral aspects of the testis, enabling it to move without twisting within the scrotum. Fluid can accumulate here (hydrocoele, see **6.8.8**).

The scrotum is supplied by branches of the femoral, internal pudendal, and inferior epigastric arteries and drains into corresponding veins. Its lymph drains to superficial inguinal nodes. Skin of the anterior part of the scrotum is innervated by the ilio-inguinal nerve (L1); its posterior part largely by scrotal branches of

the pudendal nerve (S3, 4, 5); cremaster is innervated by the genital branch of the genitofemoral nerve (L2); dartos by sympathetic fibres.

Testis and epididymis

The **testis** produces spermatozoa and also hormones which are important not only during intra-uterine development but also for the development of secondary sex characteristics at puberty and for normal reproductive function in maturity. Each testis is an oval body with an upper and lower pole (**6.8.3**). In the adult its vertical diameter is about 3 cm, its width, 2.5 cm.

Qu. 8B *Of what significance are these measurements when assessing pubertal development?*

Each testis is surrounded by a firm fibrous capsule, the **tunica albuginea** from which **septa** extend into the substance of the testis. The septa divide the testis into compartments which contain the tightly coiled **seminiferous tubules** lined with the spermatogenic epithelium, and interstitial tissue which comprises the androgen-secreting **Leydig cells**, vessels, and nerves. The U-shaped seminiferous tubules open at both ends into the **rete testis**, from which small **efferent ducts** lead to the head of the **epididymis**. The epididymis

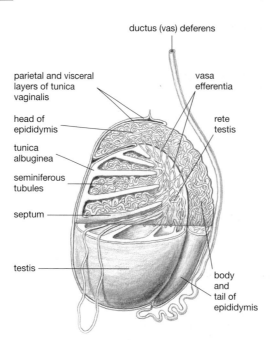

ductus (vas) deferens

parietal and visceral
layers of tunica
vaginalis

vasa
efferentia

head of
epididymis

rete
testis

tunica
albuginea

seminiferous
tubules

septum

testis

body
and
tail of
epididymis

6.8.3 Internal structure of testis and epididymis.

is a single, highly coiled, long tube attached to the posterior aspect of the testis. It has a **head, body**, and **tail**, through which the developing sperm pass, mature, and are concentrated before entering the **vas deferens**.

Blood supply, lymphatic drainage, and innervation of the testis

The arterial supply to the testis and epididymis comes from the **testicular artery** (6.7.4) which arises from the aorta at the level of the renal arteries. A limited anastomosis exists between this artery and that of the vas deferens, but the latter cannot sustain the testis if the testicular artery is damaged. Preservation of the testicular artery during operations such as vasectomy is therefore essential.

Testicular veins form a network (**pampiniform plexus**) which surrounds the cord, helping to cool the incoming arterial blood and thereby maintain a testis temperature at least 1 °C lower than that of the body core. On the right side, the testicular vein runs up the posterior abdominal wall to enter the inferior vena cava; on the left the testicular vein drains into the left renal vein. Varicosities (varicocoeles) of the pampiniform plexus are quite common, especially on the left; in such cases, the possibility that the venous drainage is blocked by a tumour involving the left kidney and renal vein must always be considered.

Lymph vessels of the testis and epididymis pass upward with the testicular vessels to drain into para-aortic lymph nodes.

The testis is **innervated** by sympathetic nerves which accompany the testicular arteries and are derived from plexuses around the renal artery and aorta. Hence, the intense autonomic activation when the testes of an adult are injured or inflamed, as can occur with mumps infection.

Vas deferens

The vas deferens, a tube of smooth muscle with a narrow lumen, is continuous with the tail of the epididymis. It passes upward along the posterior border of the epididymis, crosses the pubic tubercle where it is easy to feel, then enters the superficial ring and traverses the inguinal canal to enter the abdominal cavity at the deep inguinal ring. In this part of its course it is accompanied by:
- the testicular artery;
- the artery to the vas deferens;
- the pampiniform venous plexus;
- testicular lymphatic vessels;
- sympathetic nerves to the testis and epididymis;
- the remnants of the processus vaginalis.

These structures accompany the testis during its migration. They are surrounded by coverings derived from the abdominal wall muscles and, together, are known as the **spermatic cord**. Within the inguinal canal the spermatic cord is accompanied by the ilioinguinal nerve and the genital branch of the genitofemoral nerve.

On entering the abdominal cavity through the deep inguinal ring, the vas deferens separates from its accompanying structures and passes on to the side wall of the pelvis. Here, it runs medially and backward deep to the peritoneum to reach the base (posterior aspect) of the bladder, where it becomes expanded (**ampulla of the vas; 6.8.4**).

Seminal vesicles

Lateral to each ampulla is a sacculated pouch, the **seminal vesicle**, a single coiled tube with diverticula which produce the nutrient (fructose) vehicle for the spermatozoa. The duct of each seminal vesicle joins the ampulla of the vas deferens to form an **ejaculatory duct**, which passes forward though the prostate to open as a slit-like orifice on the crest (urethral crest) of the posterior wall of the prostatic urethra.

Prostate

The prostate (**6.8.4, 6.8.5**) is an almost spherical gland with a firm muscular stroma. It produces a part of the seminal fluid, rich in acid phosphatase, which drains via tiny ducts that open into the urethra on either side of the urethral crest.

The prostate surrounds the first (prostatic) part of the urethra, and lies on the pelvic floor attached to levator ani, which therefore lifts the prostate slightly when it contracts to raise the pelvic floor. It lies beneath the bladder, behind the pubic symphysis and in front of the ampulla of the rectum, separated only by the rectovesical fascia. Only a small part of the gland, the isthmus, lies in front of the urethra; postero-lateral to the urethra lie the **right** and **left lobes**, separated by a superficial **median furrow**. This furrow is an important feature of the normal gland when examined digitally via the rectum.

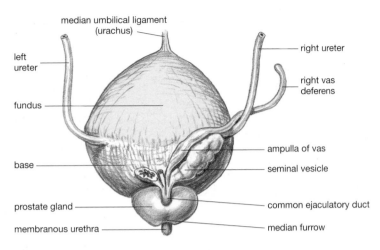

median umbilical ligament
(urachus)

left
ureter

fundus

base

prostate gland

membranous urethra

right ureter

right vas
deferens

ampulla of vas

seminal vesicle

common ejaculatory duct

median furrow

6.8.4 Posterior view of vas, seminal vesicles, prostate.

Between the ejaculatory ducts and the urethra is the small **median lobe** (6.7.6, 6.7.7) which is situated immediately beneath the trigone of the bladder. The median lobe commonly undergoes benign enlargement in elderly men. This is a common cause of the obstruction of the flow of urine during micturition; it can be detected by palpation during rectal examination because the enlargement obliterates the median furrow.

The prostate is anchored anteriorly by puboprostatic ligaments and laterally by condensations of pelvic fascia containing the prostatic venous plexus.

Qu. 8D *How could the prostate be approached surgically without opening the peritoneal cavity?*

Pea-sized **bulbo-urethral glands** (Cowper's glands; see **6.10.7**) are located on either side of the membranous urethra. They secrete a minor component of seminal fluid through ducts that pierce the perineal membrane to enter the bulb of the urethra.

Blood supply, lymphatic drainage, and innervation of the vas deferens, seminal vesicles, and prostate

The **artery to the vas deferens** is a small vessel on the wall of the vas; it arises from one of the vesical branches of the internal iliac artery. A small cremasteric artery is derived from the inferior epigastric artery. The prostate and seminal vesicles are supplied from inferior vesical and middle rectal branches of the internal iliac artery; the prostate also receives blood from the internal pudendal artery in the perineum. The **veins** of the prostate form a profuse **prostatic plexus** around the base and sides of the gland. This plexus also drains the deep dorsal vein of the penis, which reaches it by passing beneath the pubic symphysis.

Qu. 8E *To which veins does the prostatic plexus drain?*

Qu. 8F *With which veins does the prostatic plexus have a valveless communication through which prostatic cancer cells can spread?*

The **lymphatic drainage** of the prostate follows the arterial supply and venous drainage.

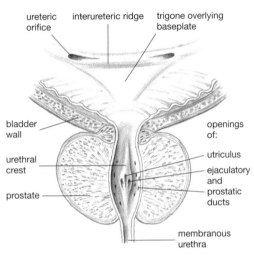

ureteric
orifice

interureteric ridge

trigone overlying
baseplate

bladder
wall

urethral
crest

prostate

openings
of:

utriculus

ejaculatory
and
prostatic
ducts

membranous
urethra

6.8.5 Base of bladder and prostatic urethra.

Qu. 8G *To which group of lymph nodes does lymph drain from:*
(a) the testis?
(b) the scrotum?
(c) the prostate and seminal vesicles?

The vas deferens, seminal vesicles, and prostate all receive a substantial sympathetic **innervation** (6.8.6) which causes contraction of the smooth muscle and emission of their contents into the prostatic urethra (see Reflexes, below).

Penis

The penis (**6.8.2**; see **6.10.6**) is formed from three masses of erectile tissue, two **corpora cavernosa** and a midline **corpus spongiosum** which surrounds the urethra. The root of the penis is situated in the perineum lying superficial to the perineal membrane. Each corpus cavernosum is attached to the inner aspect of an ischiopubic ramus, which, in males, is everted to accommodate it.

Around each corpus cavernosum is a fibro-elastic sheath and loops of muscle fibres (**ischiocavernosus**) attached to the ischiopubic ramus and the perineal membrane. Together these form the crura of the penis.

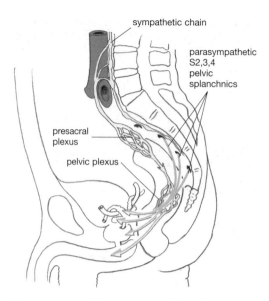

sympathetic chain

parasympathetic
S2,3,4
pelvic
splanchnics

presacral
plexus

pelvic plexus

6.8.6 Autonomic innervation of male
genital tract

The corpus spongiosum surrounds the dilated bulb of
the urethra and is itself covered by striated muscle
(**bulbospongiosus**) attached on either side to the perineal membrane and united in a midline raphe which
marks the embryonic fusion of the urogenital folds.

Vessels from the internal pudendal artery enter the
erectile tissue by piercing the perineal membrane.

Qu. 8H *What is the function of ischiocavernosus?*

In the shaft of the penis, the corpus spongiosum lies
ventral to the two corpora cavernosa. It expands distally to form the **glans penis**, the base of which is itself
expanded to form the **corona**. Surrounding the glans is
a protective fold of skin, the **prepuce** (foreskin), which
should be retractable; a small median **frenulum** passes
from its deep surface to the ventral surface of the glans.

The skin of the penis is thin and hairless and only
loosely attached to the fibrous sheaths surrounding the
erectile tissue; on the undersurface of the penis and
scrotum a median raphe marks the fusion of the two
sides during development (p. 205). Deep to the skin is
loose connective tissue devoid of fat in which lies the
superficial dorsal vein of the penis. A continuation of
the membranous layer of superficial abdominal fascia
surrounds the entire penis and separates the superficial dorsal vein from the **deep dorsal vein** and the **dorsal arteries** and **dorsal nerves**, which are terminal
branches of the internal pudendal artery and pudendal
nerve (p. 210). Surrounding and uniting the two corpora cavernosa is a layer of fibrous tissue, the **tunica
albuginea**, and within the erectile tissue are spiral
branches of the **deep arteries of the penis** derived from
the internal pudendal artery.

The penile urethra lies within the corpus spongiosum, where its lumen forms a horizontal slit; within the
glans it expands as a vertical slit, the **navicular fossa**.
The different orientation of the penile urethra and navicular fossa produces a coherent stream of urine.

Qu. 8I *To which group of nodes does lymph from the
penis drain?*

Reflex functions of the male genital tract

The testis, vas deferens, and erectile tissue receive a
dense plexus of autonomic fibres via their arteries
(**6.8.6**, see **6.10.6, 8.7**). The prostate and seminal vesicles receive autonomic fibres from the pelvic plexuses.

- **Erection** of the penis is stimulated by non-choliner-
gic, non-adrenergic autonomic nerves in the pelvic
plexuses, which innervate and dilate the arteries of
the erectile tissue. The nerves use nitric oxide as a
transmitter; Viagra amplifies their effect.
- **Emission.** Postganglionic sympathetic noradrenergic fibres stimulate contraction of the smooth muscle in the walls of the vas, seminal vesicles, and
prostate, which forces seminal fluid into the bulb of
the urethra. They also stimulate contraction of the
internal sphincter of the bladder to ensure no loss of
semen into the bladder.
- **Ejaculation.** Somatic fibres in the pudendal nerve
stimulate the rhythmic contractions of the striated
bulbospongiosus muscle which compress the bulbous urethra to cause the forcible ejaculation of
semen from the penis (see Chapter 8).

Clinical examination of the male genital system

The smooth surface of a normal testis can easily be felt
through the skin of the scrotum. The epididymis is
palpable posteriorly, and the thick, muscular vas deferens can be felt in the spermatic cord.

It may be important to determine the size of a boy's
testes at the time of puberty. This is done by comparison with a standard set of ovoid models.

The prostate and seminal vesicles can be examined
by a gloved and lubricated finger passed through the
anus into the anal canal and rectum (examination 'per
rectum'). When normal, the posterior surface of the
prostate is soft and smooth with an obvious median
sulcus, and the seminal vesicles are scarcely palpable.

Imaging the male genital system

The prostate and seminal vesicles can be visualized by
both CT (**6.8.7a**) and MRI (**6.8.7b**, see **6.3.12, 6.3.13**).
The urethra can be visualized by instillation of contrast
medium (see **6.7.9**) and contrast can also be inserted
via a fine catheter into the ejaculatory duct to outline the
seminal vesicles and vas. Cavernous erectile tissue in
the penis can also be imaged by MRI (see **6.10.12**).

The testis and epididymis are so superficial that
imaging is not often required, but the non-injurious
properties of ultrasound are used, for example, to differentiate fluid in a hydrocoele (**6.8.8**) from other
swellings of the scrotum.

Questions and answers

Qu. 8A *Of what perineal pouch is this sac a part
(see p. 208)?*

Answer The superficial perineal pouch.

Qu. 8B *Of what significance are these measurements when assessing pubertal development?*

Answer Under the influence of rising plasma testosterone levels before and during puberty, the testes gradually increase in size until they reach adult proportions. To assess pubertal development, testicular size is measured against a standard series of ovoid models.

Qu. 8C *How might one distinguish between a cyst that is connected to the lumen of the epididymis and one that is not?*

Answer Light from a pencil torch will shine through swellings filled with clear fluid, but not if the tissue contains a substantial number of spermatozoa, or is solid.

Qu. 8D *How could the prostate be approached surgically without opening the peritoneal cavity?*

Answer The prostate can be approached extraperitoneally via a suprapubic incision, opening the anterior surface of the bladder to reach the prostate.

Qu. 8E *To which veins does the prostatic plexus drain?*

Answer The prostatic plexus of veins drains into the internal iliac veins.

Qu. 8F *With which veins does the prostatic plexus have a valveless communication through which prostatic cancer cells can spread?*

Answer The vertebral veins have a valveless connection with the prostatic plexus.

Qu. 8G *To which group of lymph nodes does lymph drain from:*

(a) the testis?
(b) the scrotum?
(c) the prostate and seminal vesicles?

Answer

(a) Testicular lymph drains to para-aortic nodes via vessels running with the testicular arteries.
(b) Scrotal lymph drains to superficial inguinal nodes.
(c) Prostatic and seminal vesicle lymph drains to internal iliac nodes.

Qu. 8H *What is the function of ischiocavernosus?*

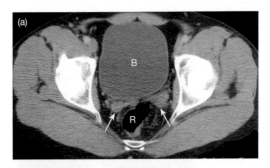

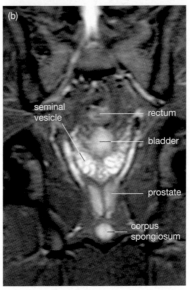

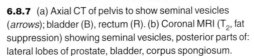

6.8.7 (a) Axial CT of pelvis to show seminal vesicles (*arrows*); bladder (B), rectum (R). (b) Coronal MRI (T$_2$, fat suppression) showing seminal vesicles, posterior parts of: lateral lobes of prostate, bladder, corpus spongiosum.

Answer Ischiocavernosus is attached to the ischiopubic ramus and covers the crus penis and, by compressing the vasodilated erectile tissue, maintains erection.

Qu. 8I *To which nodes does lymph from the penis drain?*

Answer Lymph from the skin of the penis drains to superficial inguinal lymph nodes. Lymph from the glans penis drains into a node in the femoral canal.

Qu. 8J *Of what is a patient with an enlarged prostate most likely to complain?*

Answer A patient with an enlarged prostate may complain of frequency of micturition, difficulty in emptying the bladder, a poor stream of urine, and dribbling of urine.

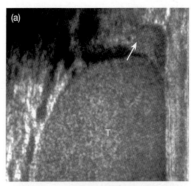

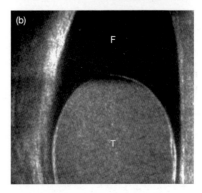

6.8.8 Ultrasound scans showing (a) normal left testis (T) and epididymis (*arrow*); (b) accumulation of fluid (F) in a hydrocoele.

Female reproductive system

Female reproductive system

The female reproductive system comprises the ovaries, the reproductive tract, and the mammary glands. The ovaries produce the female gametes. In contrast to the male, all the female gametes are produced before birth, but usually only one is released in each cycle during the reproductive period; the rest degenerate. The reproductive tract consists of the uterine (Fallopian) tubes, which convey the ova to the uterus; the uterus, in which fertilized ova develop and grow for 38 weeks (labour normally occurs at about 40 weeks after the last menstrual period) until the onset of parturition; and the vagina, through which sperm enter the female tract, the menses are shed, and the fetus is born. The mammary glands produce milk to nourish the infant after birth, and also antibodies which protect it in early life.

Development

Ovaries and female genital tract

In the female, in the absence of *SRY*, the mesodermal cells of the gonadal primordium differentiate into (pre-) granulosa cells which surround the primordial germ cells to form the primordial ovarian follicles. No androgen is produced and so the mesonephric ducts begin to disappear.

The **paramesonephric ducts** (6.9.1) form initially as an invagination of the coelomic epithelium on the lateral side of each gonadal ridge. The edges of this invagination fuse to form the duct, but the funnel-shaped rostral end of the duct opens into the peritoneal cavity near the developing ovaries. Caudally, the two paramesonephric ducts incline medially, passing anterior to the caudal extremity of the mesonephric ducts to fuse in the midline of the urorectal septum between mesonephric ducts. These induce the fused paramesonephric ducts to form the primitive uterus.

The lateral to medial course of the paramesonephric ducts raises two peritoneal folds (the right and left broad ligaments) which, with the developing uterus, form a transverse partition across the developing pelvic cavity.

The caudal end of the fused paramesonephric ducts (the **uterovaginal primordium**) contacts the dorsal wall of the urogenital sinus, raising an elevation, the **sinus** (Müllerian) **tubercle**. This proliferates to form a solid **sinuvaginal bulb**, which canalizes during the latter third of pregnancy (third trimester) to form most of the vagina (the exact contributions of different primordia to the vagina are controversial). At the junction of the developing vagina and urogenital sinus a ridge of tissue, the **hymen**, is retained; its persistence can cause discomfort at the first penetration. The lower part of the urogenital sinus creates the **vestibule** of the vagina.

The ovaries, which develop on the medial aspect of the gonadal ridges, retain this position and thus come to lie on the posterior aspect of the broad ligaments. Differential growth results in the ovary coming to lie within the pelvis. A band of connective tissue, homologous to the male gubernaculum, extends between the ovary and the labium majus. Part of the way along its length it is attached to the uterotubal junction and therefore forms two structures: the **ligament of the ovary** which connects the ovary to the uterotubal junction, and the **round ligament** of the uterus which passes from the uterotubal junction through the inguinal canal to the labium majus.

Mammary glands

The secretory epithelium of the **mammary glands** develops from the epidermis. Columns of epidermal cells proliferate and invaginate the underlying mesoderm (which will form the stroma of the gland) along two 'mammary lines', each of which extends from the axilla to the groin at about the mid-clavicular line. Normally, all but the pectoral pair of glands, which overlie the third to fifth ribs, disappear. At birth, the male and female mammary glands are indistinguishable. Development of the adult female breast commences at puberty, stimulated by oestrogen, but the gland only becomes functional after a pregnancy under the influence of pregnancy-related hormones.

Female bony pelvis

The female **bony pelvis** is adapted for the passage of the infant during parturition. It therefore differs in a number of important ways from the male bony pelvis. Examine separate male and female pelvic bones (**6.9.2a, b**).

Developmental anomalies in the female reproductive system

Because the female tract develops in the absence of gonadal hormonal signals, there are fewer anomalies than in males. Incomplete fusion of the two paramesonephric ducts can produce a uterus with two completely separate horns, a midline septum, or two cervixes. A complete or incomplete septum may also divide the vagina.

If the ureter and kidney fail to form on one side, the uterine cavity of that side can open into a vagina that does not connect with the vestibule. Remnants of mesonephric tubules and ducts can persist and, respectively, form cysts in the region of the ovary (epoöphoron and paroöphoron) or a cystic duct (Gärtner's duct) alongside the uterine tubes, uterus, and upper vagina. The presence of excess androgens during development increases the probability of mesonephric remnants persisting, and can also masculinize the external genitalia (p. 206).

The most common anomaly of breast development is the presence of accessory nipples (the 'devil's marks' of the medieval witch-hunters), or occasionally accessory mammary glands, along the 'mammary line' which runs from the normal breast position to the inguinal region.

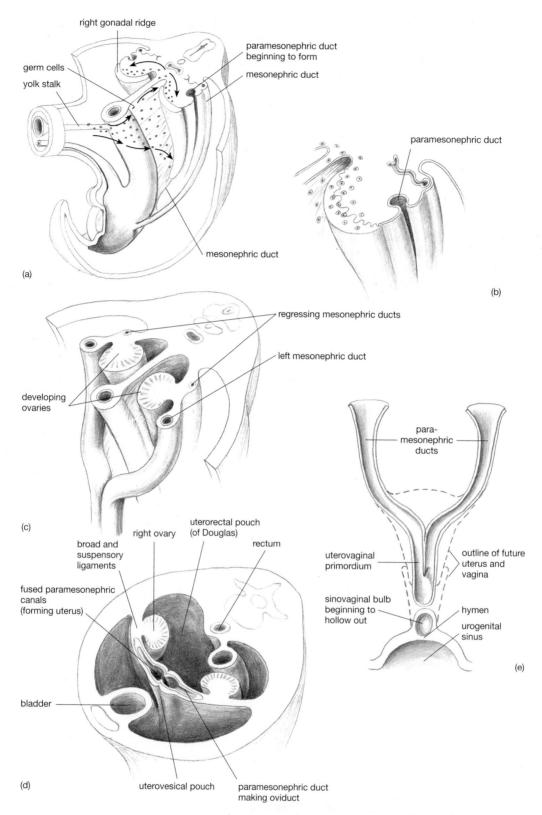

6.9.1 Development of female genital system: (a, b) migration of germ cells to intermediate mesoderm of posterior abdominal wall, where the indifferent gonad is formed; (c, d) development of ovary and paramesonephric duct system; (e) development of vagina.

Qu. 9A *What differences do you detect in:*

(a) the shape of the pelvic inlet?
(b) the shape of the iliac fossae?
(c) the shape of the subpubic arch?
(d) the dimensions of the pubic bone?
(e) the inferior pubic ramus?
(f) the shape of the greater sciatic notch?

Measure the diameter of the acetabulum in the male and female pelvis, then measure the distance between the medial edge of the acetabulum and the symphysis pubis. The ratio between the two measurements is a particularly reliable index of male and female differences.

Qu. 9B *What difference do you detect?*

Although there are characteristic features of the typical male and female pelvis, nevertheless, there is considerable variation among the pelvises of both sexes. The shape and dimensions of the pelvic inlet and outlet are of crucial importance in childbirth and therefore, particularly in a small woman, assessment of these parameters is an important part of prenatal care.

Examine the radiological anatomy of the female pelvis (see **6.1.11**), making a comparison with that of the male. Confirm the differences that you noted on the isolated bones.

Qu. 9C *Which muscle forms the floor of the pelvis?*

Peritoneal folds and spaces in the female pelvis

The peritoneum lining the female pelvis is continuous with that of the abdominal wall (**6.9.3**). Anteriorly it clothes the upper surface of the bladder. However, because the uterus lies posterior to the bladder and is normally bent forward (anteflexed) over it, the peritoneum passes on to the fundus and posterior surface of the uterus forming a shallow gutter—**the vesico-uterine pouch**—between the bladder and uterus.

Extending laterally from either side of the body of the uterus toward the side wall of the pelvis, a double sheet of peritoneum—the **broad ligament**—hangs down in front and behind the uterine (Fallopian) tubes, creating a partition running from side to side across the middle of the pelvic cavity.

From the posterior aspect of the uterus, the peritoneum continues downward to cover the upper part (posterior fornix) of the vagina and is then reflected posteriorly on to the middle part of the rectum. This creates the deepest peritoneal recess in the pelvis—the **recto-uterine pouch**, or **pouch of Douglas**. Fluid accumulating in the peritoneal cavity can collect there, so its relations are of particular importance.

Qu. 9D *How could fluid in the recto-uterine pouch be drained without damaging other areas of peritoneum?*

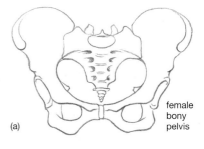

(a) female bony pelvis

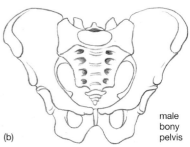

(b) male bony pelvis

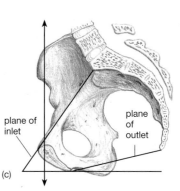

plane of inlet

plane of outlet

(c)

6.9.2 Diagram of (a) female and (b) male pelvis; (c) pelvic inlet and outlet.

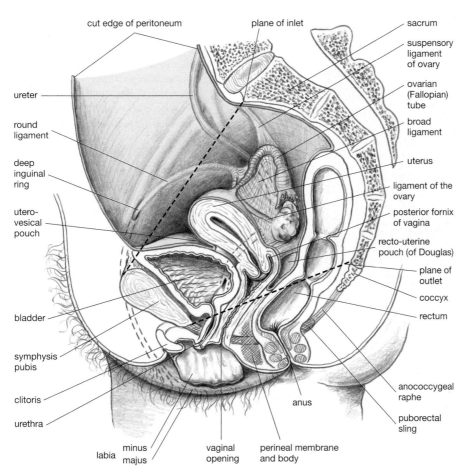

cut edge of peritoneum

plane of inlet

sacrum

suspensory ligament of ovary

ovarian (Fallopian) tube

broad ligament

uterus

ligament of the ovary

posterior fornix of vagina

recto-uterine pouch (of Douglas)

plane of outlet

coccyx

rectum

anococcygeal raphe

puborectal sling

ureter

round ligament

deep inguinal ring

utero-vesical pouch

bladder

symphysis pubis

clitoris

urethra

labia {minus majus

vaginal opening

perineal membrane and body

anus

6.9.3 Sagittal section of female pelvis. Note: the rectum is not normally distended until just before defaecation.

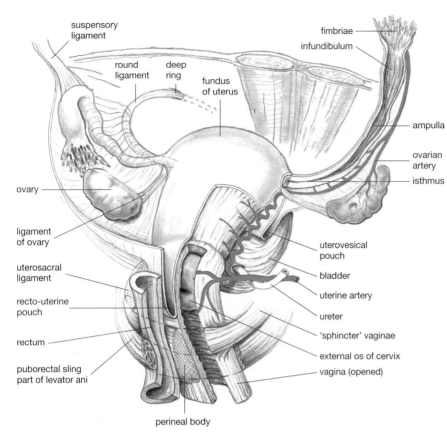

6.9.4 Posterior view of female pelvic viscera and arterial supply.

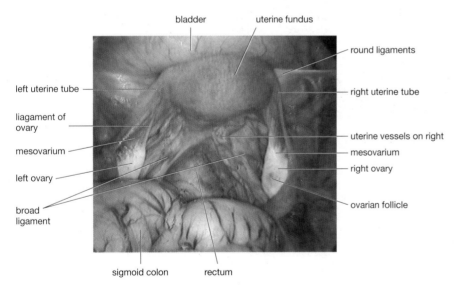

6.9.5 Laparoscopic view of female pelvis (distended with gas), showing ovaries (that on the right has a mature follicle), uterine tubes, uterus, broad ligament, ligament of ovary, round ligament, sigmoid colon, rectum, bladder.

Ovaries

The ovaries (**6.9.3–6.9.5**) lie on the posterior aspect of the broad ligament. Each is a whitish ovoid, 1–2 cm in diameter. The ovaries are covered by a layer of cuboidal epithelium, which differs from peritoneum and gives their surface a duller appearance. They are attached to

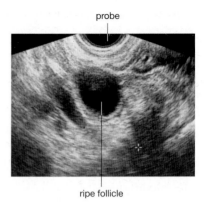

6.9.6 Ultrasound of large ovarian follicle (*arrowed*).

the broad ligament by a short pedicle of peritoneum, the **mesovarium**. They are, therefore, normally located, with the distal ends of the ovarian tubes, in the recto-uterine pouch. Here the ovaries can sometimes be palpated through the elastic posterior fornix of the vagina.

Within the layers of the broad ligament, the **ovarian ligament** extends from the ovary to the uterus immediately adjacent to the uterotubal junction. (For its continuation, the round ligament, see below.)

Each ovary consists of a cellular stroma within which are follicles at various stages of development, each of which contains an oocyte. All oocytes are formed before birth. At the beginning of a menstrual cycle (usually lasting about 28 days) each ovary is about 2 cm long and appears wrinkled. By mid-cycle one ovary is 30% larger than that on the opposite side, due to development of the dominant follicle which bulges from its surface. After ovulation and the formation of the corpus luteum, the ovary regresses in size. This cycle of events is repeated until the menopause when no more functional follicles remain and the ovaries become atrophied.

The ovaries can be visualized by laparoscopy (**6.9.5**). Pre-ovulatory follicles are large enough to be visualized by ultrasound (**6.9.6**) and ova for fertilization *in vitro* can be aspirated from the ovaries through a needle inserted through the posterior fornix of the vagina under ultrasound guidance.

Qu. 9E *What proportion of the original ova are ovulated?*

Blood supply and lymphatic drainage of the ovaries

Like the male gonadal vessels, the **ovarian arteries** originate from the aorta just below the renal arteries; the veins (which form a small pampiniform plexus) drain into either the inferior vena cava or the left renal vein. The vessels, with their accompanying sympathetic nerves and lymphatics, lie on the posterior abdominal wall, cross the pelvic brim under the peritoneum and reach the side wall of the pelvis. Here they form a small peritoneal fold—the **suspensory ligament of the ovary**—before passing between the layers of the broad ligament and mesovarium to reach the ovary, where they anastomose with terminal branches of the uterine artery. Lymphatics from the ovary, like those from the testis, drain up along the arteries to para-aortic nodes.

Female reproductive tract

Uterine tubes

The **uterine tubes** (Fallopian tubes) (**6.9.3–6.9.5**) convey ova from the ovary to the uterus. Each uterine tube extends from the ovary to the upper part of the uterus, lying within the upper margin of the broad ligament. The ovarian end of each tube opens into the peritoneal cavity close to the ovary; it is expanded (infundibulum) and its free margin is extended into a number of prolongations (fimbria) one of which is attached to the ovary.

At the time of ovulation the fimbria become more vascular and actively mobile. They almost envelop the ovary and help direct the shed ovum(a) into the infundibulum and uterine tube. Each uterine tube consists of a lateral **ampulla** and a medial, more narrow **isthmus**, which continues through the superolateral wall of the uterus to open into the upper angle of its cavity; the lumen of the opening is extremely narrow. Smooth muscle in the wall of the uterine tube helps propel ova toward the uterus.

Qu. 9F *What type of epithelium lines the uterine tubes? What is its function?*

Patency of the uterine tubes is obviously essential for reproduction. This can be examined by salpingography, in which radiopaque material is introduced into the uterus under some pressure so that it passes through the uterine cavity and uterine tubes to the peritoneal cavity. **6.9.7** is a hystero-salpingogram (*hysteros*, uterus; *salpinx*, uterine tube). Similarly, the appearance at the fimbriated end of the tube of dye introduced into the uterus can be detected by laparoscopic observation.

Uterus

The **uterus** (**6.9.3–6.9.5, 6.9.7–6.9.9**) is an inverted pear-shaped organ with thick walls of smooth muscle and a cavity lined with endometrium. It comprises a **body** and a **cervix** (**6.9.4, 6.9.8**). That part of the body which lies above the tubal openings is called the **fundus**. The body of the uterus encloses a triangular, potential cavity, the anterior and posterior walls of which are normally in contact. The uterine cavity is continuous with the cavity of the cervix via the **internal os** (mouth) of the cervix. The cylindrical cavity of the cervix opens into the vagina via the **external os**.

During the child-bearing period of life, signals from the ovary co-ordinate a menstrual cycle in which the superficial layers of the endometrium are sloughed off and regenerated about every 28 days.

The uterus in the non-pregnant state lies behind the bladder and is normally anteverted (angled forward on the vagina) and anteflexed (curved forward at the internal os) (**6.9.9**). It therefore lies over the posterior part of the upper surface of the bladder. It is very mobile and its position varies with the degree of filling of the bladder.

In post-menopausal women, and particularly in those who have had children, the uterus is often retroverted in the recto-uterine pouch (**6.9.9**).

In women who have not borne children (nulliparous) the external os is narrow and circular. After childbirth it remains relatively enlarged, forming a transverse slit (**6.9.10**). For changes in the uterus during pregnancy, see below.

Supports of the uterus and cervix

The uterus and cervix are supported in the pelvis (**6.9.11**, see **6.10.10**) not only by the orientation of the

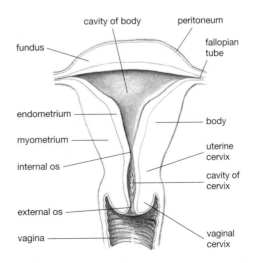

6.9.7 Hystero-salpingogram.

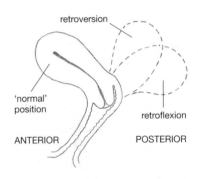

6.9.9 Common variations in the position of the uterus.

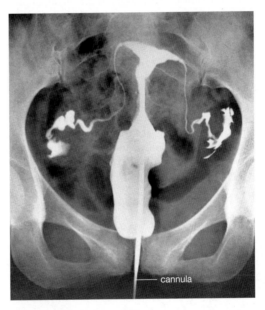

6.9.8 Coronal section of uterus.

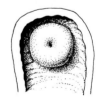

6.9.10 Cervix in nulliparous woman (above) and after childbirth (below).

bony pelvis and the presence of other pelvic organs such as the bladder, but also by:

- The **pelvic fascia**. This thick layer of fibrous tissue and smooth muscle fibres lies between the pelvic peritoneum and levator ani. Condensations of the pelvic fascia around vessels and nerves anchor and support the cervix in the pelvic floor. Condensations around each uterine artery form **lateral** or **cardinal ligaments** (of Mackenrodt) between the cervix and lateral wall of the pelvis. **Pubocervical ligaments** pass forward from the cervix to the pubic symphysis and **uterosacral ligaments** pass back around the rectum to the sacrum. These ligamentous supports of the cervix are very important in positioning the uterus.
- The **round ligaments** of the uterus help maintain the body of the uterus in an anteverted, anteflexed position. They contain some oestrogen-dependent smooth muscle, which explains the increased incidence of retroversion of the uterus after the menopause. Each round ligament extends from the uterotubal junction between the layers of the broad ligament, then runs forward to the deep inguinal ring and, accompanied by the ilioinguinal nerve, passes through the inguinal canal to the labium majus (see lymph drainage below).
- **Levator ani**, which forms the muscular floor of the pelvic cavity, provides active support. Its fibres insert into the perineal body and make a posterior sling

around and insert into the vaginal wall on which they exert a sphincter-like action.

Vagina

The **vagina** (6.9.3, 6.9.4) is a flattened muscular canal extending between the cervix and the **vestibule** of the vagina which opens on to the perineum. It lies between the urethra and bladder anteriorly, and the rectum and anal canal posteriorly. The lower (vaginal) part of the cervix projects into the anterior wall of the vagina, creating two recesses; a small **anterior fornix** and a larger **posterior fornix** of the vagina. The anterior wall of the vagina is firmly attached to the urethra, and above the urethra it is related to the trigone of the bladder.

Qu. 9G *Which fornix is closely related to the peritoneal cavity, and to what part of the peritoneal cavity is it related?*

The vestibule and clitoris will be considered with the perineum (p. 209).

Blood supply, lymphatic drainage, and innervation of the female genital tract

The **uterine artery** (6.9.4) is a branch of the internal iliac artery. Each uterine artery passes medially across

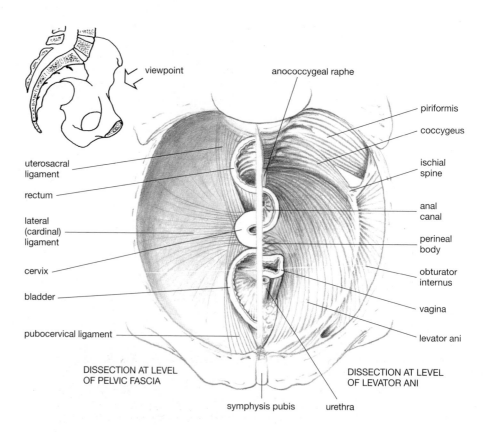

viewpoint

anococcygeal raphe

piriformis

coccygeus

ischial spine

anal canal

perineal body

obturator internus

vagina

levator ani

uterosacral ligament

rectum

lateral (cardinal) ligament

cervix

bladder

pubocervical ligament

DISSECTION AT LEVEL OF PELVIC FASCIA

symphysis pubis   urethra

DISSECTION AT LEVEL OF LEVATOR ANI

6.9.11 Diagram of two levels of the female pelvic floor, showing (on *left* at higher level) the ligaments of the uterine cervix; and (on *right* at lower level) fibres of levator ani.

the pelvic floor in the base of the broad ligament, then crosses over the ureter to reach the sides of the cervix just above its lateral ligaments. The uterine arteries supply the upper part of the vagina, then run upward along the sides of the uterus and along the uterine tubes within the layers of the broad ligament. There they anastomose with terminal branches of the ovarian arteries.

Qu. 9H *To what important structure is the uterine artery closely related at the side of the cervix?*

The uterine and ovarian arteries anastomose in the broad ligament and this anastomosis supplies the uterine tube. Enormous enlargement of the vasculature occurs during pregnancy.

Veins from the uterus and uterine tubes follow the arteries, respectively, to the internal iliac vein and the ovarian vein. **Lymph** from the uterus passes largely to internal iliac nodes. (Communications with superficial inguinal nodes are formed by lymphatics running with the round ligaments so that, rarely, secondary deposits from a malignant tumour of the uterus can appear in the inguinal nodes.) Lymph from the ovary, like that from the testis, drains upward with the ovarian arteries to para-aortic nodes around the renal arteries.

The female reproductive tract is **innervated** by sympathetic nerves in the pelvic plexus running with the ovarian vessels, and by pelvic parasympathetic nerves. Pain from the body of the uterus is conveyed along sympathetic fibre tracts via the presacral plexus to upper lumbar segments of the cord. Nerves sensing distension of the cervix (which is rather insensitive to pain) run with pelvic parasympathetic nerves to the sacral cord.

Sensory innervation of the cervix forms the afferent limb of the reflex by which dilatation of the cervix during parturition leads to secretion of oxytocin from the posterior pituitary to increase uterine contractions.

Clinical examination of the female reproductive tract

No components of the female genital system other than a pregnant or otherwise enlarged uterus or a large ovarian cyst can be detected by simple palpation through the anterior abdominal wall.

The vagina and cervix can be examined with gloved fingers inserted via the introitus and also by inspection if the vagina is held open by an instrument (speculum).

Qu. 9I *What structures would you expect to be able to palpate during a vaginal examination?*

The uterus is best examined bimanually. One hand on the suprapubic anterior abdominal wall palpates the uterine fundus, the other examines digitally via the vagina. In this way uterine size and anterversion/retroversion can be assessed; also whether it is mobile or fixed, tender or non-tender. The walls of the vagina are normally very elastic and the ovaries may be palpable through the posterior fornix, particularly when enlarged.

The width of the bony pelvis can be assessed during vaginal examination because the ischial spines may be palpable through the posterior fornix. During childbirth it may be necessary to perform a vaginal examination to determine the position of the fetal head (the anterior fontanelle is usually palpable).

Contents of the recto-uterine pouch can also be palpated through the posterior vaginal fornix.

Qu. 9J *What structures might you expect to find in the recto-uterine pouch?*

Imaging the female reproductive tract and fetus

Qu. 9K *What unexpected feature do you see in radiograph 6.9.12?*

Simple radiology does not reveal the tract, but the patency of the uterine tubes can be revealed by a salpingogram (**6.9.7**). MRI—either median sagittal (**6.9.13a**), coronal (see **6.5.16b**), or axial sections (**6.9.13b, c**)—can show the ovaries, uterus, and vagina, as well as the recto-uterine pouch. In **6.9.13b** an ovarian cyst is present.

Qu. 9L *What abnormality is clearly visible in MRI 6.9.14?*

Ultrasound scans and **magnetic resonance imaging** are the preferred methods for examining the fetus, because they do not involve the use of ionizing radiation. Ultrasound scans can be used to visualize the fetal skeleton, heart, brain, kidneys, bladder, and limbs.

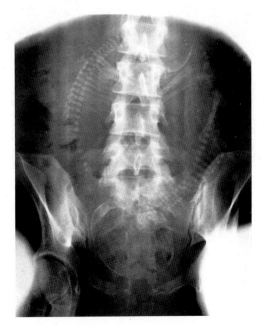

6.9.12 See Qu. 9J.

Cervical smear

Samples of cells scraped from the surface of the cervix (cervical smear test) are taken from women to check for changes in the cells which would suggest the development of cancer of the cervix.

Placental villus sampling

It is sometimes necessary to take a biopsy from the placenta to determine if genetic abnormalities are present in the fetal tissue. This is done by inserting an instrument through the cervix via the vagina.

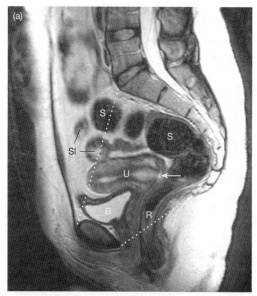

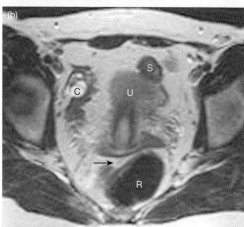

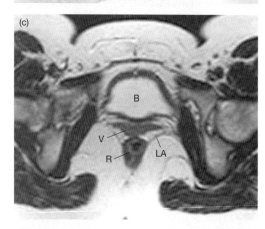

6.9.13 MRIs of the female pelvis. (a) midline sagittal section. Normal uterus (U), cervix (*arrow*), vagina. The planes of the inlet and outlet are marked. (b) Axial section. Anteverted body of the uterus (U) overlies the bladder; fat underlying the recto-uterine pouch (*arrow*) separates the rectum (R) from the posterior fornix of the vagina; ovarian cyst (C). (c) Lower axial section. The vagina (V) lies between rectum and full bladder (B). Levator ani (LA) is the muscle sheet separating the pelvis from fat of ischiorectal fossa.

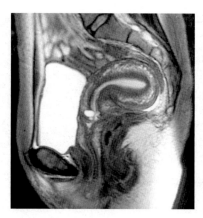

6.9.14 See Qu. 9K.

when ova are removed from mature follicles (**6.9.6**) for fertilization *in vitro*.

The uterus, uterine tubes, and ovaries, like many other abdominal organs, can be examined visually by means of an endoscope inserted into the peritoneal cavity through either the anterior abdominal wall (**6.9.5**) or the posterior fornix of the vagina (laparoscopy). This procedure is used for diagnosis, and for sterilization by ligation of the uterine tubes.

Changes in pregnancy and parturition

Growth of the uterus during pregnancy

During pregnancy, which normally ends about 40 weeks after the last menstruation, the uterus grows and enlarges to accommodate the developing fetus and placenta. Its upper border reaches the level of the xiphoid process at 36 weeks of pregnancy (**6.9.17**), but toward term is usually a little lower because the fetal head descends into the pelvis.

It is the body and fundus of the uterus that undergo the main expansion; but the upper third of the cervical canal also expands during and after the second month of pregnancy, to form the 'lower uterine segment'. After expulsion of the fetus and placenta, the uterus normally regresses in size to almost its prepregnant state within a few weeks.

Palpation of the anterior abdominal wall is used to determine the upper extent of a uterus enlarged by pregnancy. Parts of the fetus, such as the head, can be felt after about 24 weeks of pregnancy and movements of the fetal limbs after about 28 weeks.

Soft-tissue changes at the end of pregnancy

To permit the passage of the head and body of the fetus through the cervix, vagina, and introitus considerable changes have to occur in the tissues that make up the walls of these parts of the reproductive tract. These are brought about primarily by prostaglandins secreted by the placenta.

6.9.15 shows, in a 22-week pregnancy, fetal head, spine, and thorax. The MRI (**6.9.16**) of a 34-week pregnancy provides a median sagittal sectional view on which can be seen parts of the fetus, uterus, and placenta.

Ultrasound is also used to determine the size of ovarian follicles and to guide the aspiration procedure

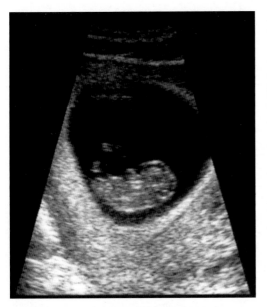

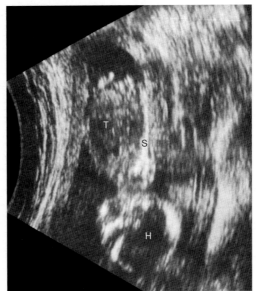

6.9.15 Ultrasound scans of pregnancies at (a) 11 weeks and (b) 22 weeks. Fetal thorax (T), spine (S), and head (H).

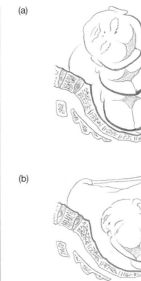

(a)

(b)

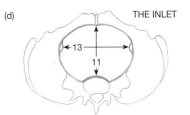

(c)

(d) THE INLET

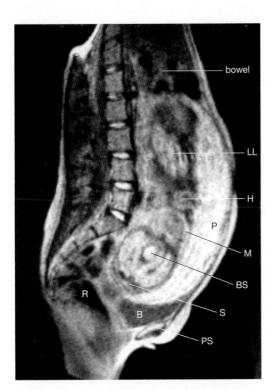

6.9.16 MRI (sagittal section) of 34-week pregnant woman. Fetal lower limb (LL), hand (H), brainstem (BS) and surrounding brain, skull (S), maternal bowel, rectum (R), bladder (B), pubic symphysis (PS), mandible (M).

The cervix, which has to be competent during pregnancy to retain the fetus and its amniotic sac, undergoes considerable changes in the glycosaminoglycans of its connective tissues, to become sufficiently distensible to allow the passage of the fetal head. Similar, but less marked, changes occur in the vagina and introitus.

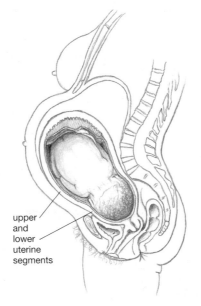

6.9.17 Sagittal section of pregnant woman near term.

Passage of the fetus through the pelvis and perineum during childbirth

Toward the end of pregnancy, the ligaments of the pelvis also become more relaxed due to the action of hormones secreted by the placenta.

A day or two before childbirth commences, the head of the fetus, which has been lying within the uterus above the level of the pelvic inlet, usually descends to the level of the pelvic inlet where it becomes 'engaged' (**6.9.18**). The pelvic inlet is widest in transverse dimension and so the head of the fetus, which is longer than it is broad, is orientated transversely.

As the head is squeezed through the pelvis during labour, the occiput of the infant is usually the first part of the head to contact the pelvic floor. This slopes

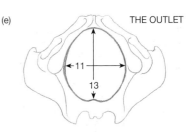

(e) THE OUTLET

6.9.18 (a)–(c) Rotation of fetal head during its passage through the female pelvis; with dimensions (cm) of (d) inlet and (e) outlet of typical (gynaecoid) female pelvis.

downward and forward at an angle of about 15° (see **6.9.2c**). The slope of the pelvic floor then directs the head of the fetus to rotate so that the occiput faces anteriorly. The head of the fetus is then appropriately placed to pass through the bony pelvic outlet, which is wider antero-posteriorly than transversely.

The process of birth is usually more difficult if the occiput faces posteriorly or if the buttocks rather than the head enters the pelvis (breech-first presentation).

Normally the changes in the soft tissues permit the passage of the fetus without tearing, but sometimes tears of the perineum occur. Unless such tears are carefully controlled, they can involve the anal canal (p. 210). To prevent uncontrolled tearing an incision (episiotomy) may be made to enlarge the vaginal opening.

Mammary glands (breasts)

At birth each breast is rudimentary. It comprises little more than few relatively unbranched ducts in a connective tissue stroma. These are confined to the area immediately under the pinkish pigmented **areola** which surrounds the **nipple**. The structure of the gland remains rudimentary until puberty when, in the female, concentrations of plasma oestrogen increase and fat is laid down around the growing duct system; the alveoli remain primitive.

If pregnancy occurs, increasing concentrations of oestrogen and progesterone are responsible respectively for proliferation of the ducts and of the secretory alveoli around the terminal ducts.

After parturition, production of milk is stimulated primarily by prolactin from the anterior pituitary gland. Ejection of milk from the breast is caused by a reflex in which nerves stimulated by suckling cause the secretion of oxytocin from the posterior pituitary. The oxytocin causes the contraction of specialized myoepithelial cells which surround the alveoli, expressing milk into the ducts. The suckling stimulus also maintains the secretion of prolactin.

In the male, mammary tissue remains rudimentary throughout life, but will respond to hormones produced in the body or administered therapeutically.

Examination of the living breast

Each mammary gland (**6.9.19**) lies on the chest wall largely over pectoralis major. Its form and size are very variable, but its base is more constant, extending from the 2nd to the 6th ribs between the sternum and the anterior axillary line.

The base of a normal breast is mobile on the underlying chest wall, although some pockets of breast tissue can pierce the deep fascia. Any tethering to deep structures can signify disease. An **axillary tail** (see **6.9.21**) extends upward and laterally into the axilla. In the male, the nipple usually lies in the fourth intercostal space in the mid-clavicular line; in adult females its position is necessarily more variable.

When examining the breasts, their contours should be studied with the arm in a variety of positions to detect any irregularity or asymmetry. As the arm is moved, the breasts should move freely on the chest wall and the overlying skin should not be tethered. The nipples are normally everted. The pigmentation of the areola varies, becoming darker in the lactating breast (**6.9.20**).

Palpation of the breast is done initially with the palm of the hand and not with the fingertips, which sense too many completely normal irregularities in the breast tissue.

Structure of the breast

The breast of a normal adult female (**6.9.21**) consists of 15–20 **lobules** of glandular tissue arranged radially around the nipple, separated from one another by

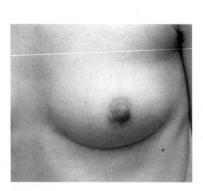

6.9.19 Non-lactating breast.

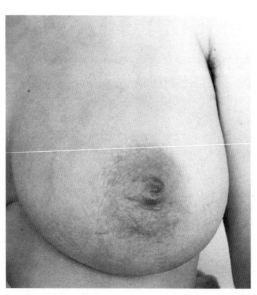

6.9.20 Lactating breast.

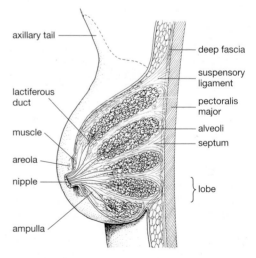

6.9.21 Components of the breast and nipple.

fibrous tissue and supported by fat. The lobules of glandular tissue lie in the superficial fascia, although the axillary tail may extend deep to the pectoral fascia. Each lobule is drained by a **lactiferous duct**, which is dilated to form an ampulla before opening on to the skin of the raised nipple. The nipple is surrounded by circular muscle fibres which erect it during suckling.

Areolar glands open around the nipple. They secrete an oily sebaceous protective lubricant and are most prominent in pregnancy and lactation.

The breasts are supported by fibrous **suspensory ligaments** which extend from the deep fascia overlying pectoralis major to the dermis. Other fibres pass to the skin from the connective tissue of the breast. If breast tissue is abnormally swollen with tissue fluid these fibres pull in the skin ('peau d'orange' appearance). After the menopause the mammary tissue atrophies. The breasts tend to sag because the 'suspensory ligaments' are dependent on oestrogen. Oestrogen replacement therapy can reduce this change.

Qu. 9M *What changes occur in the female breast during the life cycle and to what, in general, can these changes be attributed?*

Blood supply, lymphatic drainage, and innervation of the breast

The **arterial** supply and venous drainage of the breast come from two main sources: the internal thoracic (internal mammary) artery and its intercostal branches medially; and branches from the axillary artery laterally. These vessels enlarge considerably during lactation. The veins correspond to the arteries. It is noteworthy that the intercostal veins communicate directly with vertebral veins and that secondary malignant deposits in the vertebrae are a common complication of breast cancer.

The **lymphatic drainage** of the breast is extensive and of considerable importance in breast cancer (**6.9.22** and see Vol. 1, p. 105). About 75% of the lymphatics from breast tissue drain into clusters of **axillary nodes** lying inferior to, behind, and above (apical nodes) pectoralis minor. The remaining lymphatics drain mostly the medial aspect of the breast and pass into nodes lying within the chest around the internal thoracic artery in the second to fourth intercostal spaces. These nodes communicate freely across the midline behind the sternum.

Some lymph also drains to infraclavicular nodes, to nodes on the opposite side of the chest or downward into nodes within the sheath of rectus abdominis. Lymph from an axillary tail of the breast often drains into nodes around the subscapular artery on the posterior wall of the axilla.

The innervation of the breast is from the segmental intercostal nerves (p. 117).

Imaging the breast

The lower margins of the breasts are usually visible as curved shadows on plain A-P radiographs of the chest or abdomen (see **5.4.7**). Breast soft tissues are visible

on CTs and MRIs of the chest, and MRI is increasingly used for diagnosis of breast pathology, particularly in premenopausal women in whom breast tissue is radiologically very dense.

To obtain sufficient definition, radiographic examination of a breast (a mammogram) involves compressing each breast individually between the X-ray source and the imaging plate. Mammograms are

---

Breast lumps

Many different conditions can cause irregularities or 'lumps' in the breast tissue. Some are small cysts or benign tumours, which can enlarge but do not invade other tissues. Malignant transformation (breast cancer) is, however, a major cause of death among women.

Because malignant tumours invade surrounding tissues, if a breast lump is discovered, any attachment to the skin or chest wall should be determined. Attachment to the chest wall can be detected by asking the patient to move the relevant arm, in particular to medially rotate the shoulder against resistance in order to contract pectoralis major which underlies the breast.

The draining lymph nodes should also be examined to detect any enlargement or attachment (affected nodes are usually not tender), bearing in mind that the lymph drainage may be atypical if tumour deposits have blocked some lymph vessels.

Involvement of the internal thoracic nodes can allow cancer cells to spread to the pleura, causing an effusion of fluid into the pleural cavity.

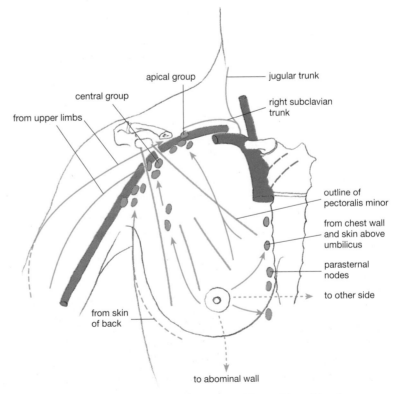

6.9.22 Lymphatic drainage of breast to axillary and parasternal (internal thoracic) nodes.

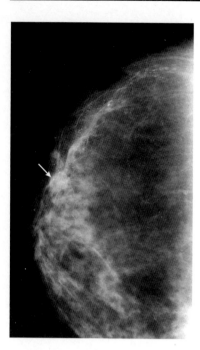

6.9.23 Mammogram: showing a carcinoma with microcalcification.

routinely used for screening in some breast clinics. Mammogram **6.9.23** shows a breast carcinoma (arrow) which distorts the local architecture and contains characteristic microcalcification.

Questions and answers

Qu. 9A *What differences do you detect in:*

(a) *the shape of the pelvic inlet?*
(b) *the shape of the iliac fossae?*
(c) *the shape of the subpubic arch?*
(d) *the dimensions of the pubic bone?*
(e) *the inferior pubic ramus?*
(f) *the shape of the greater sciatic notch?*

Answer

(a) the pelvic inlet in the female is transversely oval; in the male it is heart-shaped because the sacral promontory bulges forward;
(b) the iliac fossa in the female is broader and shorter than in the male;
(c) the subpubic arch is wider and more rounded ('Roman') than in the male ('Norman');
(d) the pubic bone is proportionally larger in the female;
(e) the inferior pubic ramus of the male is everted and is thickened by the attachment of ischiocavernosus;
(f) the greater sciatic notch is wider in the female.

Qu. 9B *What difference do you detect?*

Answer You should expect to find that, in the male, the width of the acetabulum and distance from the acetabulum to the symphysis are comparable. However, in the female, a broader pelvis is combined with a somewhat smaller acetabulum, making the former substantially larger than the latter.

Qu. 9C *Which muscle forms the floor of the pelvis?*

Answer The pelvic floor is formed by levator ani. It slopes downward and forward by about 15° to the horizontal (see **6.9.2c**).

Qu. 9D *How could fluid in the recto-uterine pouch be drained without damaging other areas of peritoneum?*

Answer By insertion of a draining needle through the posterior fornix of the vagina.

Qu. 9E *What proportion of the original ova are ovulated?*

Answer Approximately 1 million ova are present at birth, of which about 500 will be ovulated in a 38-year reproductive span (a proportion of about 1:2000).

Qu. 9F *What type of epithelium lines the uterine tubes? What is its function?*

Answer The uterine tubes are lined with a ciliated, secretory epithelium which transports the ova toward the uterus and nourishes the conceptus.

Qu. 9G *Which fornix is closely related to the peritoneal cavity, and to what part of the peritoneal cavity is it related?*

Answer The posterior fornix of the vagina is covered with peritoneum which is reflected on to the anterior surface of the middle part of the rectum, thus forming the deepest recess of the peritoneal cavity, the recto-vaginal pouch (pouch of Douglas).

Qu. 9H *To what important structure is the uterine artery closely related at the side of the cervix?*

Answer The uterine artery runs with the ureter on the pelvic floor in the base of the broad ligament; the artery then crosses anterior to the ureter before ascending between the two layers of the broad ligament.

Qu. 9I *What structures would you expect to be able to palpate during a vaginal examination?*

Answer The cervix, laterally the ovaries, posteriorly the rectum. Laterally the ischial spines may be palpable, which enables one to assess the width of the pelvis.

Qu. 9J *What structures might you expect to find in the recto-uterine pouch?*

Answer The ovaries; a loop of sigmoid colon or small bowel is commonly present. Occasionally a long appendix extends into the pouch.

Qu. 9K *What unexpected feature do you see in radiograph **6.9.12**?*

Answer Twin fetuses are present in the uterus of this pregnant woman.

Qu. 9L *What abnormality is clearly visible in MRI **6.9.14**?*

Answer The uterus is retroverted. The bladder is very distended, and a fluid filled cyst is present in the anterior wall of the uterine cervix.

Qu. 9M *What changes occur in the female breast during the life cycle and to what, in general, can these changes be attributed?*

Answer In children only a rudimentary duct system is present. During puberty the duct system grows and rudimentary alveoli appear, but the principal increase in breast size is due to deposition of fat. During pregnancy both the ducts and the alveoli develop and, at the end of pregnancy, secretion of milk by the alveoli commences and continues during lactation. After suckling ceases, apoptosis of the glandular tissue occurs, macrophages invade breast tissue, and the breast returns to near the pre-pregnant state. After the menopause, the reduction in oestrogen causes diminution both of the breast tissue and of the support provided by the dependent suspensory tissue.

Perineum

Perineum

The perineum is the diamond-shaped area beneath the pelvic outlet and between the two lower limbs. It transmits the output tracts of the two principal excretory systems (urinary and faecal) and contains the external genitalia. It is firmly attached to the pelvic floor. A line drawn transversely in front of the two ischial tuberosities divides the area into an anterior (urogenital) triangle, which is sexually differentiated and divided horizontally by a perineal membrane, and a posterior (anal) triangle which contains the anal canal and its sphincters. In the male, the penile urethra is long and surrounded by erectile tissue; it provides a channel through which urine is voided and semen is ejaculated. The testes in their scrotal sac are slung beneath the urogenital triangle. In the female the urethra is short and opens into the vestibule just in front of the vagina. The vaginal orifice is surrounded anteriorly and laterally by erectile tissue which extends forward beneath the labia minora to form the clitoris.

Development of male and female external genitalia

In the fourth week of fetal life in both male and female fetuses a **genital tubercle** develops at the anterior end of the cloacal membrane while **urogenital folds** and **labioscrotal swellings** appear on either side (**6.10.1**). At the same time, within the pelvis, the **urorectal septum** grows down to fuse with the interior aspect of the cloacal membrane. This fusion divides the cloacal membrane into an **anal membrane** (posteriorly) and a **urogenital membrane** (anteriorly). When these membranes perforate to form the anal and urogenital orifices, the indifferent stage of development is at an end.

In a 10–12-week **male** fetus, under the influence of androgens, the genital tubercle grows forward, and in so doing pulls the urogenital folds forward. The resulting **urethral groove**, which is formed between the urogenital folds on the ventral surface of the phallus, is lined with cloacal endoderm. At the tip (**glans**) of the developing penis a cord-like ingrowth of ectoderm develops, which canalizes to become continuous with the urethral groove. At the same time, the urogenital folds fuse in the midline to convert the urethral groove into the **penile** (**spongy**) part of the **urethra**. If this fusion is incomplete, the urethra opens on to the ventral surface of the penis, a condition known as **hypospadias**.

The erectile tissues and their associated muscles form from mesenchymal tissue within the penis. Androgens also stimulate the labioscrotal folds to fuse in the midline, forming two scrotal pockets into which, on either side, the processus vaginalis and testis, with its accompanying vessels and nerves, will descend.

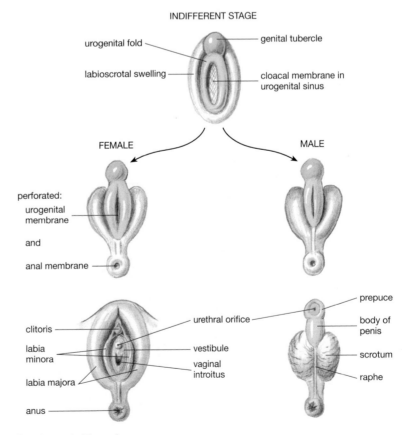

6.10.1 Development of the perineum.

During the twelfth week of fetal life a circular ingrowth of ectoderm forms around the glans. This ingrowth breaks down so that, after birth, the foreskin (**prepuce**) is usually retractable.

In a **female** fetus, in the absence of androgens, the genital tubercle (phallus) develops more slowly to form the small **clitoris**. The urogenital folds do not fuse in the midline but remain as the **labia minora**. The labioscrotal folds also remain unfused and form two large folds, the **labia majora**. The anterior part of the urogenital sinus, on to which both the urethra and the vagina open, is known as the **vestibule**.

For the development of the anal canal, see p. 145; for the development of the vagina, see p. 192.

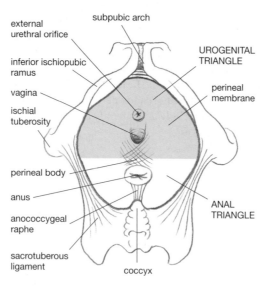

6.10.2 Bony and ligamentous boundaries of the perineum.

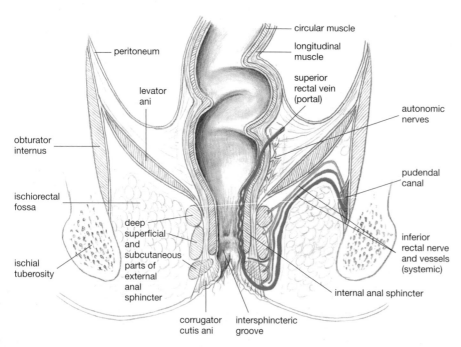

6.10.3 Coronal section through rectum and anal canal.

Perineum

The perineum overlies the outlet of the pelvis and is the diamond-shaped area bounded by the subpubic arch, the ischiopubic rami, the sacrotuberous ligaments, and the coccyx (**6.10.2**). It extends between the skin and the floor of the pelvic cavity, which is formed by levator ani. Levator ani therefore forms the sloping roof and medial wall of deeper compartments of the perineum. The lateral wall of the perineum is formed by the lower part of the obturator internus muscle attached to the obturator membrane, and the ischiopubic ramus.

The perineum can be divided into an anterior **urogenital triangle** and a posterior **anal triangle** by a line drawn transversely in front of the two ischial tuberosities (**6.10.2**). The anal triangle is similar in both sexes; the urogenital triangle is markedly different.

The urogenital triangle is subdivided into superficial and deep compartments by the **perineal membrane** (**6.10.2**, **6.10.8**, **6.10.9**), a tough, triangular horizontal sheet of fibrous tissue that spans between the ischiopubic rami. In the midline of its posterior margin, the perineal membrane and levator ani unite to form the fibrous **perineal body**.

The anal canal passes through levator ani to emerge into the anal triangle; the urethra and the vagina pierce both levator ani and the perineal membrane to enter the urogenital triangle.

Anal triangle

The anal triangle (**6.10.2**, **6.10.6**, **6.10.9**) lies between the posterior border of the perineal membrane anteriorly and the ischial tuberosities, sacrotuberous ligament, and coccyx that form its postero-lateral margin. It is overlapped posteriorly by the lower border of gluteus maximus.

It contains the midline **anal canal** and, on either side, the **ischiorectal fossae**. Anterior to the anal canal is the **perineal body**; posteriorly, between the anal canal and the coccyx, is the **anococcygeal raphe**. Levator ani slopes downward and medially from the inner aspect of the pelvic bone to these midline structures. It is strongly attached to the perineal body, forms a posterior sling round the anorectal junction, and its interlacing fibres form the raphe.

The **anal canal** (**6.10.3**) is a sphincteric tube, 3–4 cm in length, which passes downward and backward from the lower part of the rectum (ampulla) to the **anus**, a skin-lined aperture lying about 4 cm in front of the tip of the coccyx.

The mucosal lining of the anal canal reflects its composite origin (p. 145). Its upper part is lined with columnar epithelium like that of the rectum; there is a transitional zone which overlies the internal venous plexus; and the lower part is lined with keratinized stratified squamous epithelium (skin). The upper border of the skin is known as Hilton's white line.

The acute angulation at the anorectal junction is maintained by the forward pull of fibres of levator ani (puborectalis) which form a sling posterior to the bowel. The anal canal provides a passage for defaecation and its sphincters prevent faecal material or flatus from leaving the rectum until such time as is convenient.

Defaecation

For most of the time the rectum is empty and the sphincter mechanism (the anal sphincters and puborectalis) tonically contracted. Peristaltic movement of faeces from the sigmoid and descending colon into the rectum causes the desire to defaecate. When the rectum is sufficiently distended, the internal anal sphincter and puborectal sling relax (rectosphincteric reflex) and the defaecation reflex is initiated. Faeces are expelled by contraction of the rectal smooth muscle, aided, if necessary, by contraction of abdominal wall muscles, which increases intra-abdominal pressure (see p. 114; **8.6**).

During childhood, humans gradually develop the ability to restrain the reflex by contraction of the voluntary external sphincter until defaecation is socially convenient.

The walls of the rectum and anal canal contain numerous mechanoreceptors which can distinguish the presence of gas, liquid, and solid in the lumen. The internal anal sphincter, which is normally tonically contracted, relaxes intermittently to equalize pressure in the rectum and anal canal, thereby facilitating this 'anorectal sampling'. One cause of faecal incontinence is chronic constipation and impaction of faeces, which damages the nerve endings and prevents effective sampling.

Faecal continence

Continence is maintained by the tonic contraction of **anal sphincters** (**6.10.3**) which surround the canal and by the **anal cushions** (**6.10.4**), three pads of mucosa and vascular submucosa, the volume of which can be varied to seal the canal internally.

- The **internal anal sphincter** is the downward continuation of the circular smooth muscle of the rectum; it surrounds the upper three-quarters of the anal canal and is responsible for 80% of the sphincter action. Outside this, the longitudinal muscle coat of the rectum is replaced by fibro-elastic tissue, which forms a sleeve around the anal canal. Muscular fibres of levator ani insert into this sleeve. Distally, the fibrous sleeve splits into bundles that, with a few voluntary subcutaneous muscle fibres (corrugator cutis ani), pucker the perianal skin to complete the act of defaecation.

Qu. 10A *Stimulation of which nerves would cause the internal anal sphincter to relax?*

- The **external anal sphincter** is the voluntary striated muscle sphincter. It acts as a whole but is described as having three continuous parts. The deep part is a thick ring of muscle which blends with the puborectal sling of levator ani at the recto-anal junction; on rectal examination it can be felt as an **anorectal ring**. The superficial part also lies outside the internal

sphincter and comprises fibres which mostly run between the coccyx and the perineal body. The subcutaneous part lies distal to the internal sphincter, the interval between them forming the **intersphincteric groove**, which is palpable on rectal examination.

- The **anal cushions** are three masses of the thick, very vascular submucosa in the anal canal. Numerous arterio-venous anastomoses allow the volume of the cavernous tissue to be controlled so that it fills the anal canal. The cushions are situated around the anal circumference at 3, 7, and 11 o'clock (viewed in the lithotomy position; **6.10.5**). The cushions are more or less separated by vertical folds of mucosa (anal columns). **Anal glands** secrete mucus into **anal sinuses** at the base of these columns. The anal cushions provide a spongy variable volume 'washer' on which the sphincters can contract to provide a waterproof seal. The cushions are supported by bundles of smooth muscle derived from the internal anal sphincter and which pass through the cavernous tissue to attach to the mucosa below the pectinate line. During defaecation these fibres contract to flatten the anal cushions against the internal sphincter.

Ischiorectal fossa

Lying laterally on either side of the anal canal is a wedge-shaped potential space, the **ischiorectal fossa** (**6.10.6**). It is bounded above and medially by levator ani, and laterally by obturator internus, the ischial tuberosities, sacrotuberous ligament, and attached gluteus maximus. Posteriorly the two ischiorectal fossae communicate via the small space behind the anal canal; anteriorly each is continuous with the narrow space which extends forward between the perineal membrane and levator ani to the posterior aspect of the pubis. The terminal parts of the pudendal nerves and vessels pass in this space, then exit between the pubis and perineal membrane to reach the dorsum of the penis or clitoris (**6.10.2**).

Qu. 10B *Which other vessel passes through this space, and how does it terminate?*

The ischiorectal fossae are filled with **fat** which is fluid at body temperature. This supports the rectum and anal canal and permits their expansion during defaecation.

Blood supply, lymphatic drainage, and innervation of the anal canal

The mucous membrane of the upper part of the anal canal (which develops from the hind gut) is supplied

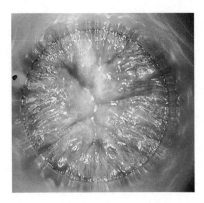

6.10.4 Anal cushions, viewed from above.

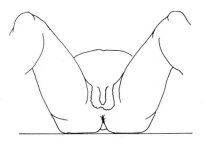

6.10.5 Lithotomy position.

Haemorrhoids

Haemorrhoids ('piles') is a common condition that is probably due to chronic constipation. The shearing force of hard faeces disrupts the anal cushions and causes them to prolapse downward through the anus. The mucosa covering the haemorrhoids is very fragile, so fresh blood on the faeces is common.

Anal varicose veins

Anal varicose veins are very rare and quite different from haemorrhoids. They are caused by back pressure in the hepatic portal venous system. They are not situated in the position of the anal cushions but form dilated veins deep to the skin of the anal orifice.

Anal fissures

Anal fissures are tears in the mucous membrane caused by hard faecal material. Infection can spread through the fissures into the ischiorectal fossa. Such fissures are very painful because of the rich somatic sensory nerve supply of the anal canal.

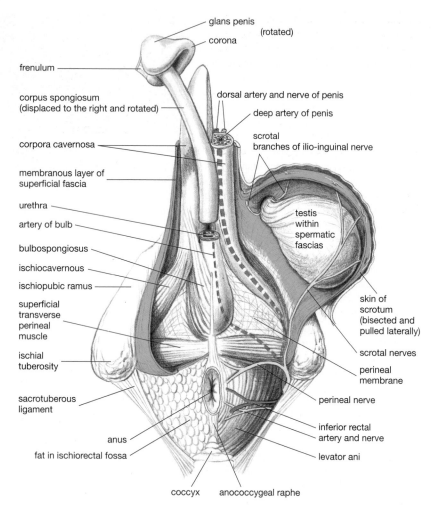

frenulum

corpus spongiosum
(displaced to the right and rotated)

corpora cavernosa

membranous layer of
superficial fascia

urethra

artery of bulb

bulbospongiosus

ischiocavernous

ischiopubic ramus

superficial
transverse
perineal
muscle

ischial
tuberosity

sacrotuberous
ligament

anus

fat in ischiorectal fossa

glans penis
(rotated)

corona

dorsal artery and nerve of penis

deep artery of penis

scrotal
branches of ilio-inguinal nerve

testis
within
spermatic
fascias

skin of
scrotum
(bisected and
pulled laterally)

scrotal nerves

perineal
membrane

perineal nerve

inferior rectal
artery and nerve

levator ani

coccyx   anococcygeal raphe

6.10.6 Perineum in the male.

by the **superior rectal artery** (the terminal branch of
the inferior mesenteric artery). The **middle rectal
artery** (from the internal iliac) supplies the muscular
coat of the upper part of the anal canal, anastomosing
via arterioles with the mucosal supply. The **inferior
rectal artery** (a branch of the pudendal artery) is badly
named because it does not supply the rectum, but supplies all layers of the lower part of the anal canal.

Veins from the anal canal drain via superior rectal
veins to the hepatic portal system and via middle and
inferior rectal veins to the (systemic) internal iliac
veins. The anal canal is therefore a site of portal–systemic venous anastomosis (see the clinical box, Anal
varicose veins).

Lymphatics from the anal canal run with the blood
vessels to nodes around iliac vessels in the pelvis and to
superficial inguinal nodes.

The **innervation** of the upper part of the anal canal
is provided by the enteric plexus, and the internal
sphincter by sympathetic nerves. The lower part of the
anal canal and the voluntary external anal sphincter
are supplied by inferior rectal branches of the
(somatic) pudendal nerve.

Urogenital triangle in the male

Spanning the interval between the two ischiopubic
rami is a tough triangular sheet of fibrous tissue, the
perineal membrane (6.10.6). This provides attachment for the **root of the penis** which is formed from
three masses of erectile tissue: the two **corpora cavernosa**, and the **corpus spongiosum** which encloses the
urethra. The very narrow **membranous urethra**
pierces the perineal membrane just in front of its free
posterior border (see **6.7.7**) then expands to form the
bulb of the urethra within the corpus spongiosum.
The two **corpora cavernosa** lie laterally on the perineal
membrane attached to the inferior pubic rami which
are everted to accommodate them.

Around each corpus cavernosum is a fibro-elastic
sheath and loops of muscle fibres (ischiocavernosus)
attached to the ischiopubic ramus and the perineal
membrane. Together these form the crura of the penis.

Likewise, the corpus spongiosum containing
the bulb of the urethra is covered by muscle fibres (bulbospongiosus) which are attached on either side to the
perineal membrane and unite in a midline raphe
which marks the embryonic fusion of the urogenital
swellings to enclose the penile urethra.

Qu. 10D *What are the functions of bulbospongiosus in
the male?*

The two corpora cavernosa and the corpus spongiosum unite anteriorly to form the body of the penis.
This is suspended from the pubic symphysis by a suspensory ligament and enclosed in a loose tube-shaped
extension of the membranous layer of the superficial
fascia of the anterior abdominal wall and by hairless
skin. The two corpora cavernosa lie dorsally, fused in
the midline; the corpus spongiosum containing the
urethra lies ventrally and expands distally to form a
cap—**glans penis**—which itself is expanded to form
the **corona** over the ends of the corpora cavernosa.
Covering the glans is a retractable fold of skin the
prepuce (foreskin) attached by a small median septum
(frenulum) to the base of the glans. The space between
the glans and foreskin is lubricated by sebaceous
glands.

The penile urethra which, in the main part of the
corpus spongiosum forms a horizontal slit, dilates
within the glans, to form a vertical slit—the navicular
fossa. This change in orientation of the lumen produces a coherent stream of urine.

Superficial to the perineal membrane is a space—
the **superficial perineal pouch (6.10.6, 6.10.7**, see
7.6)—which contains not only the root of the penis
and its muscles, but also vessels, and nerves and, on
each side, the testis, epididymis, and vas partially
enclosed in the tunica vaginalis. The pouch is formed
by a continuation of the membranous layer of superficial fascia of the anterior abdominal wall (p. 113;
6.10.8) which passes around the penis and scrotum and
into the perineum, where it is attached laterally to
the ischiopubic rami and is fused posteriorly to the

posterior border of the perineal membrane and the fibromuscular perineal body.

Qu. 10E *A young boy slips while climbing a tree and lands astride a lower branch. On examination there is bruising of the perineum and a swelling which distends the scrotum and penis and the lower abdominal wall. What is likely to have happened?*

To the perineal body are attached:
- muscular fibres of levator ani;
- the superficial part of the external anal sphincter;
- bulbospongiosus;
- superficial and deep transverse perineal muscles which lie on either surface of the perineal membrane and are also attached to the ischiopubic rami.

Deep to the perineal membrane is a potential space which is continuous with the ischiorectal fossa posteriorly, and with the superficial perineal pouch anteriorly through a narrow gap between the perineal membrane and pubic symphysis. The space is roofed by the undersurface of levator ani, fibres of which pass laterally around the undersurface of the prostate ('levator prostatae') to attach to the perineal body. Between the perineal membrane and the pelvic floor is the **membranous urethra** surrounded by the **external (voluntary) urethral sphincter**, which is partly blended with the deep transverse perineal muscles. On each side of the membranous urethra lie the **bulbo-urethral glands**. Their mucous secretions are carried by ducts which pierce the perineal membrane and enter the bulb of the urethra, providing lubrication during sexual intercourse.

Qu. 10F *Does the prostate lie above or below levator ani?*

Urogenital triangle in the female
(6.10.9, 6.10.10)

The **labia majora** (derived from the unfused lateral genital folds) are prominent fat-filled cutaneous folds which extend downward and backward from the mons pubis to form the lateral boundaries of the pudendal cleft.

Qu. 10G *Which ligament of the uterus ends in the labium majus?*

Between the labia majora are two smaller cutaneous folds devoid of fat—the **labia minora** (derived from the medial genital swellings)—which form the lateral boundaries of the **vestibule**. Anteriorly, the labia minora are joined together, enclosing the clitoris and forming the anterior boundary of the vestibule.

The attachments of the membranous layer of superficial fascia (**6.10.9**) and the perineal membrane are the same as in the male. In the female, however, there is no enclosed superficial perineal pouch. Also the vagina pierces the perineal membrane between the perineal body and the urethra, and the urethra and vagina open directly on to the skin of the vestibule. The perineal membrane is therefore a much less distinct layer in the female than in the male.

The perineal muscles in the female are similar to those in the male. The transverse perineal muscles are identical; each ischiocavernosus covers a **crus of the clitoris** (the equivalent of the corpus cavernosum in

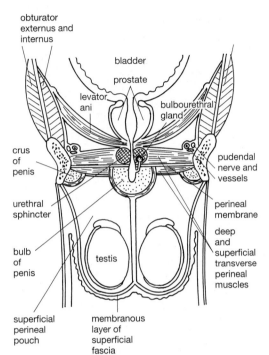

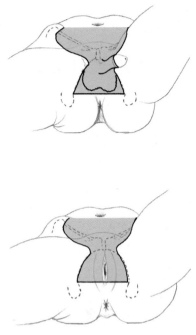

6.10.8 Attachments of membranous layer of superficial fascia.

6.10.7 Schematic diagram of coronal section of male pelvis and perineum, showing structures lying superficial and deep to the perineal membrane. The bulbo-urethral gland is shown on one side only.

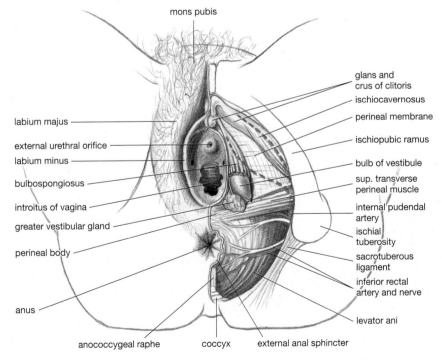

6.10.9 Perineum in female.

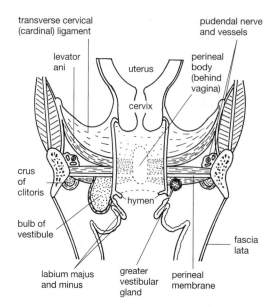

transverse cervical
(cardinal) ligament

pudendal nerve
and vessels

levator
ani

uterus

perineal
body
(behind
vagina)

cervix

crus
of
clitoris

hymen

bulb of
vestibule

fascia
lata

labium majus
and minus

greater
vestibular
gland

perineal
membrane

6.10.10 Coronal section of female pelvis
and perineum.

ramus (i.e. the lateral wall of the ischiorectal fossa) and
continues forward in the space deep to the perineal
membrane. Its terminal branches emerge through a
gap at the anterior border of the membrane on to the
dorsum of the penis or clitoris (**6.10.3, 6.10.6**).

Branches of the nerve and artery leave the canal to
supply the various components of the perineum.

- **Inferior rectal vessels** and **nerves** (**6.10.3, 6.10.6**)
  pass medially across the roof of the ischiorectal fossa
  then pass down to supply the anal canal and surrounding skin.
- **Perineal vessels** and **nerves** are distributed to the
  skin of the posterior part of the labia majora or scrotum, to the penile urethra, and to the muscles of the
  perineum.
- The **deep artery to the penis/clitoris** and **artery of the
  bulb** pierce the perineal membrane to enter, respectively, the corpora cavernosa and spongiosum. Those
  entering the corpora cavernosa form spiral (helicine)
  vessels. These arteries open into the cavernous spaces
  and their dilation allows the increased blood flow
  which is responsible for the expansion of the cavernous tissue during erection (see p. 186).
- The **dorsal nerve** and **artery of the penis** (**6.10.6**) (or
  clitoris) run forward deep to the perineal membrane
  then emerge beneath the pubic symphysis on to the
  dorsum of the penis or clitoris, along which they
  pass, deep to the membranous layer of superficial
  fascia.

Qu. 10H *What nerve, and what dermatome supplies the
anterior parts of the scrotum or labia majora?*

Qu. 10I *If local anaesthesia was required for (a) enlargement of the vaginal outlet during childbirth (episiotomy), or (b) to incise a peri-anal abscess, at what site
would you inject local anaesthetic?*

The **venous** drainage of the cavernous tissue passes
largely via a **deep dorsal vein of the penis** to the prostatic venous plexus and thence to the internal iliac
veins. A midline **superficial dorsal vein** drains the skin
and superficial fascia of the penis to the saphenous
vein in the upper part of either the left or right thigh.

Lymph from the perineum passes partly with the
internal pudendal artery to the internal iliac nodes.
Anteriorly, the skin and fascia of the penis and scrotum
drains to superficial inguinal nodes. The glans of the
penis is said to drain to an inguinal node located in the
femoral canal.

Qu. 10J *A patient has a painful lump in the groin which
appears to be an inflamed superficial inguinal node. In
what areas would you look for the primary infection?*

Clinical examination of the perineum

Clinical examination of the rectum

A lubricated, gloved finger can be used to examine
the anal canal, the lower part of the rectum, and any

Damage to the perineum during childbirth

The fetal head is large and, even though
the vaginal outlet is very elastic at the
end of pregnancy, tears can occur
during childbirth and can even extend to
involve the anal musculature (**6.10.11a**).
To prevent this, the outlet may have to
be extended by an incision (episiotomy)
which must avoid the perineal body. Any
damage that occurs during childbirth
must be carefully repaired, since failure
to do so can lead to faecal incontinence
if the anal sphincters are damaged, and
weakness of the pelvic floor if the
perineal body is damaged. Weakness of
the pelvic floor can lead to a tendency of
the uterus to prolapse, and consequent
distortions of the urethra and rectum
which can cause urinary or faecal
incontinence.

Postnatal exercises include
voluntary contractions of levator ani,
the fibres of which lie close to the
lateral wall of the vagina before
inserting into the perineal body. These
fibres support the vagina and have
sometimes been referred to as
'sphincter vaginae' (see **6.9.11**).

---

the male); the **bulbospongiosus** muscles lie around the
lower end of the vagina, beneath the labia minora of
either side. They cover the **bulbs of the vestibule**, two
large masses of erectile tissue (equivalent to the corpus
spongiosum), the anterior ends of which unite to form
the **glans of the clitoris.**

The body of the **clitoris** is therefore formed from a
glans and two crura but, unlike the body of the penis, it
does not enclose the urethra. A small suspensory ligament attaches the clitoris to the inferior border of the
pubic symphysis, but the clitoris lies almost entirely
within the folds of the labia minora.

Deep to the perineal membrane, the urethra and
external (voluntary) urethral sphincter are firmly
attached to the anterior wall of the vagina. Hence, urinary incontinence is common when the vaginal wall is
distorted by a prolapse of the uterus (p. 196).

The greater **vestibular glands** in the female (homologues of the bulbo-urethral glands in the male) lie
one on each side of the vaginal orifice under cover
of the erectile tissue of the labia majora. Their ducts
open on to the epithelial surface between the hymen
and the labium minus to provide lubrication for
intercourse.

The **perineal body**, with its attachments both to the
pelvic floor via levator ani and to the muscles and
fascia of the perineum, is even more crucial to the
anatomical and functional integrity of the perineum
in the female than in the male.

Blood supply, lymphatic drainage, and innervation of the perineum

The **internal pudendal artery** and **pudendal nerve** (S2,
3, 4) leave the pelvis through the greater sciatic notch
below piriformis, and enter the perineum by hooking
around the ischial spine. The neurovascular bundle
runs forward in a fascial canal (**pudendal canal**) on the
inner surface of the ischial tuberosity and ischiopubic

structures in the peritoneal pouch anterior to the rectum. The rectum is normally empty of faeces. Within the anal canal, the intersphincteric groove and the muscular ring formed by the puborectal sling and deep part of the external sphincter should be palpable. Postero-laterally, in both sexes one can feel the coccyx, ischial spines, and tuberosities.

In the male: anteriorly, the posterior aspect of the prostate should be soft, with a distinct median furrow between the two lateral lobes; below this the penile bulb and membranous urethra can be felt, particularly if a urethral catheter is in place. The seminal vesicles are palpable only if enlarged.

In the female: the cervix can be felt through the anterior rectal wall and the extent of its dilation during parturition determined.

Clinical examination of the vagina

A lubricated, gloved finger can also be used to examine the vagina and cervix. During this procedure, the ovaries and other contents of the recto-uterine pouch may be felt through the elastic posterior fornix of the vagina. The cervix can also be examined visually if the introitus is held open with a speculum. The size, shape, and position of the uterus is determined bimanually (p. 197).

Imaging

Radiological techniques used to investigate the lower alimentary (p. 154), urinary (p. 176), and reproductive tracts (p. 186 & 197) have been covered.

Ultrasound is also used to investigate the integrity of the lower rectum, anal canal, and other features of the pelvis (endosonography; see Chapter 4). Such ultrasound examinations allow the detection of tears of the anal sphincters (**6.10.11a**) and an assessment of the extent to which tumours have infiltrated through the wall of the bowel (**6.10.11b**).

MRI, T_2-weighted to show fluid as bright, reveals the cavernous erectile tissue in the perineum (**6.10.12**).

Questions and answers

Qu. 10A *Stimulation of which nerves would cause the internal anal sphincter to relax?*

Answer Stimulation of the pelvic parasympathetic nerves relaxes the smooth muscle of the internal anal sphincter as part of the defaecation reflex. At the same time it stimulates the other rectal smooth muscle to expel the faeces.

Qu. 10B *Which other vessel passes through this space, and how does it terminate?*

Answer The deep dorsal vein of the penis passes between the subpubic arch and the perineal membrane to drain into the prostatic plexus.

Qu. 10C *Where might abscesses of the ischiorectal fossa discharge ('point')?*

Answer Abscesses in the ischiorectal fossa might point on to peri-anal skin or into the anal canal.

Qu. 10D *What are the functions of bulbospongiosus in the male?*

Answer Bulbospongiosus is attached to the perineal membrane and encompasses the corpus spongiosum and the bulb of the urethra. Its contraction empties the bulb after micturition and its rhythmical contraction causes the ejaculation of semen during orgasm.

Qu. 10E *A young boy slips while climbing a tree and lands astride a lower branch. On examination there is bruising of the perineum and a swelling which distends the scrotum and penis and the lower abdominal wall. What is likely to have happened?*

Answer The extravasated urine would pass into the superficial perineal pouch and thence into the anterior abdominal wall deep to the membranous layer of superficial fascia. Because the membranous layer is attached tightly to the fascia lata of the thighs, the pubic arches, and the posterior margin of the perineal membrane, urine cannot pass into the thighs or the anal triangle.

Qu. 10F *Does the prostate lie above or below levator ani?*

Answer The prostate lies above levator ani and therefore on the pelvic floor.

Qu. 10G *Which ligament of the uterus ends in the labium majus?*

Answer The round ligament of the uterus—a derivative of the gubernaculum.

Qu. 10H *What nerve, and what dermatome supplies the anterior parts of the scrotum or labia majora?*

Answer The anterior part of the scrotum/labium majus is supplied by the ilio-inguinal nerve (L1).

Qu. 10I *If local anaesthesia was required for (a) enlargement of the vaginal outlet during childbirth (episiotomy), or (b) to incise a peri-anal abscess, at what site would you inject local anaesthetic?*

Answer (a) For episiotomy, local anaesthetic is infiltrated all round the vaginal orifice. (b) To incise a peri-anal abscess, the pudendal nerve can be blocked close to where it passes around the ischial spine.

Qu. 10J *A patient has a painful lump in the groin which appears to be an inflamed superficial inguinal node. In what areas would you look for the primary infection?*

Answer The infection might be in the perianal region, the penis, the lower limb, or in the lower abdominal wall, all of which drain to inguinal lymph nodes.

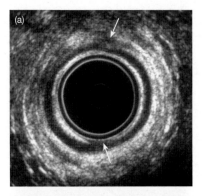

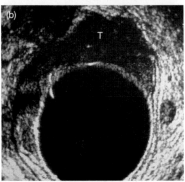

6.10.11 Ultrasound images of abnormalities of rectum; compare with **6.5.17**. (a) Tears of the internal and external anal sphincters (*arrows*); (b) a tumour (T) infiltrating the rectal wall.

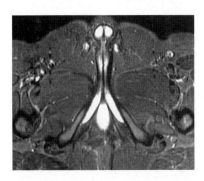

6.10.12 Axial T_2 MRI of male perineum. The cavernous erectile tissue appears bright.

Anatomy of the abdomen revealed by CT and MRI

Anatomy of the abdomen revealed by CT and MRI

Clinical diagnosis is increasingly being aided by visualization of the internal anatomy of the abdomen by CT and by MRI. One advantage of MRI is that the parameters for acquisition of the images can be varied to highlight different tissues. **6.11.1b–j** (g–j from a different individual) is a series of axial CTs through the normal abdomen and pelvis at various levels (**6.11.1a**). As with all axial sectional images, the view is from beneath the section (i.e. as if you are looking at the section from the feet of the body). Increasingly, however, sections in other planes (coronal, sagittal, oblique), calculated to show a particular structure to the best advantage, are being used. **6.11.2a–d** are coronal MRIs through progressively more posterior sections of the abdomen.

The anatomical understanding that you have acquired from three-dimensional sources, such as living anatomy and prosections, must mentally be translated into sections through the body in order to identify the structures revealed by the images. It is important that students of medicine should be able to recognize on these images the normal range of shape, position, and relations of all the large abdominal and pelvic organs, blood vessels, etc. If you can do this, you will have a good three-dimensional mental image of the structure of the body. To aid such learning, it is a good idea to place a cover across the adjacent labelled diagrams and discover how many structures you can recognize before checking your identifications.

Other CT and MRI images of particular organs are: 6.3.11–6.3.13; 6.4.16; 6.5.16; 6.6.10; 6.8.7; 6.9.13, 6.9.14, 6.9.16, 6.1012, 6.11.3.

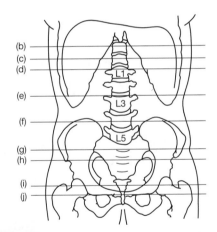

(a)

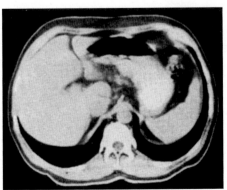

(b)

left lobe of liver

caudate lobe of liver

portal vein

right lobe of liver

inferior vena cava

ribs

lung in right costo-diaphragmatic recess

gas in stomach

splenic flexure of colon

spleen

aorta

right and left crura of diaphragm

T11

(b)

6.11.1 *Continued.*

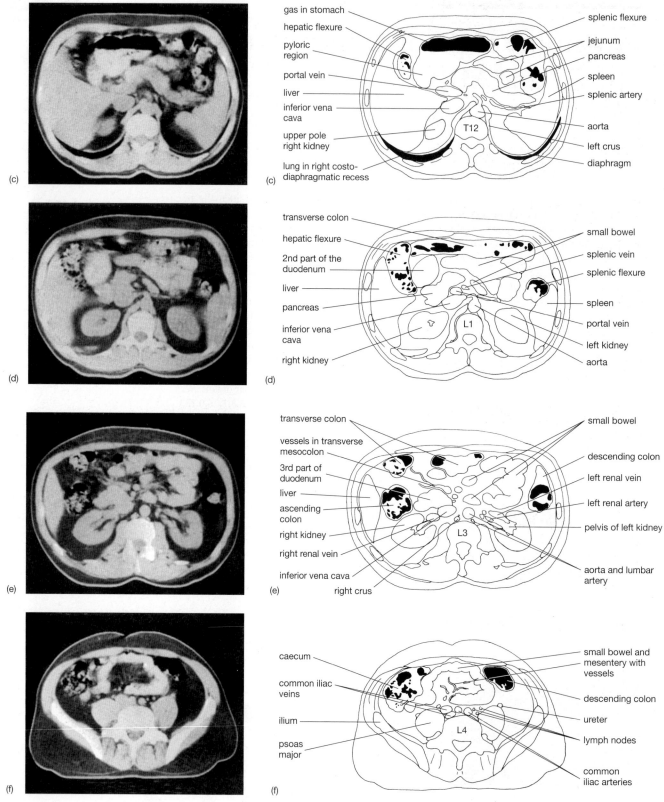

(c)

gas in stomach
hepatic flexure
pyloric region
portal vein
liver
inferior vena cava
upper pole right kidney
lung in right costo-diaphragmatic recess

splenic flexure
jejunum
pancreas
spleen
splenic artery
aorta
left crus
diaphragm

T12

(c)

(d)

transverse colon
hepatic flexure
2nd part of the duodenum
liver
pancreas
inferior vena cava
right kidney

small bowel
splenic vein
splenic flexure
spleen
portal vein
left kidney
aorta

L1

(d)

(e)

transverse colon
vessels in transverse mesocolon
3rd part of duodenum
liver
ascending colon
right kidney
right renal vein
inferior vena cava
right crus

small bowel
descending colon
left renal vein
left renal artery
pelvis of left kidney
aorta and lumbar artery

L3

(e)

(f)

caecum
common iliac veins
ilium
psoas major

small bowel and mesentery with vessels
descending colon
ureter
lymph nodes
common iliac arteries

L4

(f)

6.11.1 *Continued.*

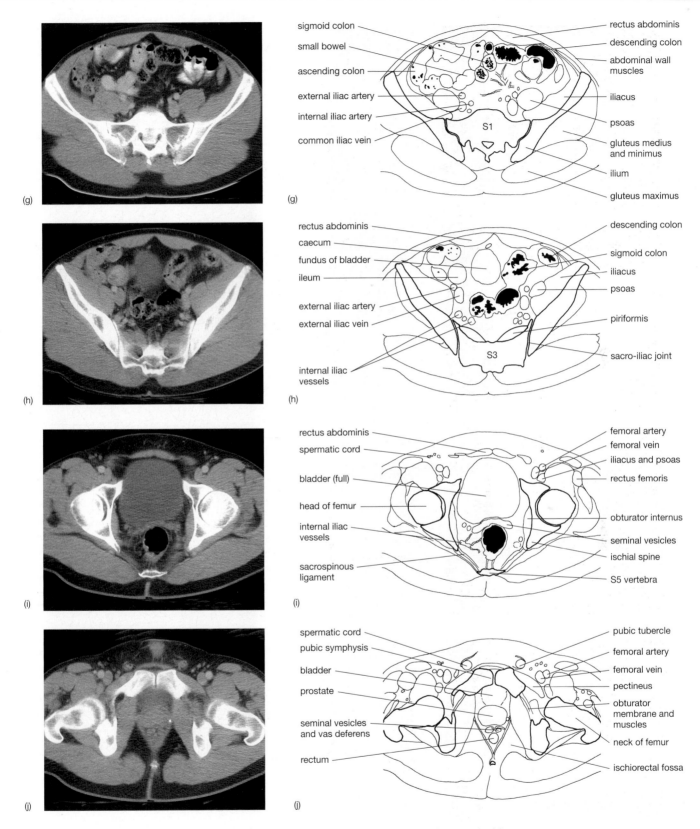

6.11.1 (a) Levels of sections (b)–(j) through the abdomen and pelvis. (b)–(j) Axial CTs through male abdomen and pelvis.

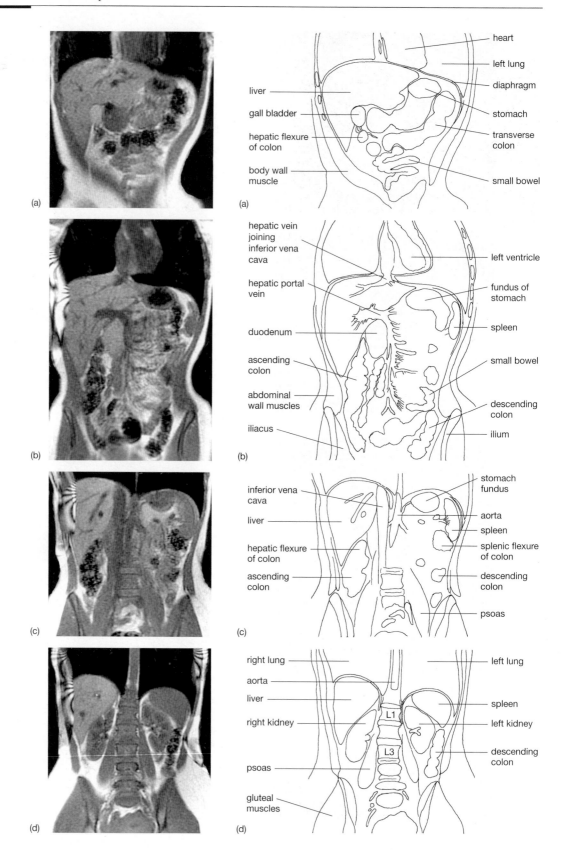

6.11.2 Coronal MRIs through abdomen and pelvis.

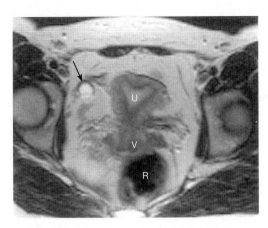

6.11.3 Axial MRI of female pelvis. U, Uterus; V, vagina; R, rectum, ovarian cyst (arrow).

Endocrine glands and tissues of the thorax and abdomen

Endocrine glands and tissues of the thorax and abdomen

Endocrine glands are ductless glands composed of endocrine cells which secrete hormones into the bloodstream and surrounding tissues. Some endocrine cells, such as those of the gut, are not collected together into glands but form diffuse endocrine tissues. Hormones entering the bloodstream affect distant organs via the endocrine route; hormones can also act locally on surrounding cells by a paracrine action or on the hormone-producing cells themselves by an autocrine action. All endocrine tissues have a profuse blood supply. The main endocrine glands are shown in **7.1**.

Endocrine glands of the thorax

Thymus

The **thymus** is a lymphoid organ situated in the anterior mediastinum (p. 83). Its endocrine cells secrete a variety of different peptide hormones which affect the maturation of lymphocytes in the thymus and in the periphery.

Respiratory tract

The epithelium of the **respiratory tract** contains isolated endocrine cells scattered throughout its mucosa.

7.1 Endocrine glands of the body.

These produce peptides and amines, but their functions are largely unknown.

Heart

The **heart** produces **atrial natriuretic peptide** in myocytes in the walls of the atria. It is released when the atria are distended, and causes loss of sodium in the urine and inhibition of renin release.

Qu. A *Aberrant endocrine tissue derived from endocrine glands normally located in the neck may also be found in the anterior mediastinum. Which glands, and why does this occur?*

Endocrine glands of the abdomen

Within the **abdomen** are a number of endocrine tissues of different types: the paired **adrenal** (**suprarenal**) glands are entirely endocrine in function; the pancreas contains clusters of endocrine cells—the **islets of Langerhans**—which form an endocrine organ within the exocrine tissue; the **gonads** produce hormones in both their interstitial cells and germinal epithelium; the **kidneys** produce and metabolize a number of hormones; and the **alimentary tract** contains a diffuse endocrine system within its epithelium which plays a large part in co-ordinating the digestive process.

Adrenal (suprarenal) glands: cortex & medulla, fetal adrenal (see also p. 176; 6.7.4, 6.7.10)

The adrenal glands are paired endocrine organs (see also p. 176). The right adrenal is situated on the upper pole of the right kidney, lying between the diaphragm and the bare area of the liver, just lateral to the inferior vena cava. The left adrenal is situated over the medial aspect of the upper pole of the left kidney, lying on the left crus of the diaphragm and coeliac plexus behind the

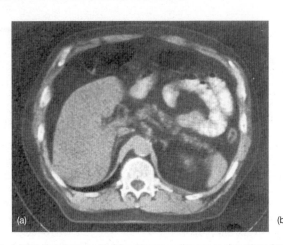

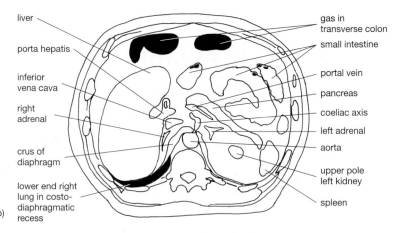

liver

porta hepatis

inferior
vena cava

right
adrenal

crus of
diaphragm

lower end right
lung in costo-
diaphragmatic
recess

gas in
transverse colon

small intestine

portal vein

pancreas

coeliac axis

left adrenal

aorta

upper pole
left kidney

spleen

7.2   (a) CT of abdomen with (b) explanatory diagram to show the adrenal glands.

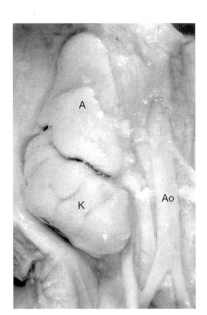

7.3   Fetal adrenal (A) and kidney (K),
aorta (A$_o$).

lesser sac and stomach. The adrenals have a characteristic appearance on axial CT (**7.2**).

Each adrenal comprises an outer **cortex** and an
inner **medulla**. The cortex is derived from the intermediate column of mesoderm on the posterior wall of
the embryo; the medullary cells are derived from the
neural crest.

The cortex consists of three distinct zones of cells all
of which secrete **steroid hormones**. The outer zona
glomerulosa secretes the mineralocorticoid aldosterone; the zona fasciculata secretes the glucocorticoid
cortisol; and the inner zona reticularis secretes androgenic sex steroids.

The cells of the adrenal medulla secrete the **catecholamines** adrenaline and noradrenaline. The juxta-
position of the cortex and medulla is necessary
because conversion of noradrenaline to adrenaline
requires a glucocorticoid-dependent enzyme.

The fetal adrenal

In humans, the fetal adrenal (**7.3**) is relatively very large
due to the presence of an additional fetal zone of the
cortex. This produces an androgen (dehy-
droepiandrosterone sulphate; DHEA-S) which is the
essential substrate for the production of oestradiol by
the placenta (which lacks the enzymes necessary to
convert pregnenolone to androgens). After birth the
fetal zone of the adrenal rapidly involutes and the
adrenals adopt their definitive pattern of cortical zones.

Blood supply, lymphatic drainage, and innervation

Like all other endocrine glands the adrenals are well
supplied with arterial blood. Indeed, during stress, the
vessels dilate rather than constrict like the rest of the
splanchnic vessels. They receive branches directly from
the aorta, and also from the renal arteries and phrenic
arteries. They drain via a single, wide adrenal vein
which leaves the hilus of the gland to enter the inferior
vena cava on the right and the renal vein on the left.
Lymph from the adrenals drains to para-aortic glands
and thence to the cisterna chyli.

Preganglionic sympathetic (thoracic splanchnic)
nerves pass through the crura of the diaphragm to
synapse on cells of the adrenal medulla. These cells are
the homologues of the ganglion cells of the sympathetic system but release their amine products directly
into the bloodstream rather than from nerve terminals.
Postganglionic sympathetic fibres control the adrenal
blood vessels, which dilate during stress.

Qu. B *From where do suprarenal medullary cells derive
preganglionic fibres? What is the principal neurotransmitter at the synapse between preganglionic fibres and
the adrenal medullary cells?*

Islets of Langerhans (see also p. 165)

The islets of Langerhans (about 1 million in number)
are aggregates of endocrine cells shaped like small
irregular balls and scattered throughout the pancreas.
They are derived from endodermal cells which grow
out from the epithelium lining the pancreatic ducts.
Islets in the body and tail of the pancreas secrete
insulin from their beta cells, glucagon from their alpha
cells, and somatostatin from their delta cells; all these
hormones have profound effects on carbohydrate, fat,
and protein metabolism. Islets in the uncinate process
of the pancreas produce mostly pancreatic polypeptide, the function of which is unclear. Individual islets
are too small to visualize by MRI, but endocrine
tumours in the pancreas (see clinical box) can be
detected by MRI if they reach a sufficient size.

Blood vessels and innervation

Each islet has its own small arteriole derived from the
arteries to the adjacent exocrine pancreas. The hormone-
rich venous blood drains via the hepatic portal vein
first to the liver, which is the principal site of action of
the islet hormones and which therefore receives the
hormones at a much higher concentration than the
peripheral tissues.

The pancreas receives postganglionic sympathetic
and preganglionic parasympathetic vagal innervation

which acts on both the exocrine and endocrine tissues. Vagal cholinergic stimulation enhances the secretion of both insulin and glucagon, sympathetic catecholamines enhance glucagon secretion but inhibit insulin secretion. By these actions, when vagal tone is high after a meal metabolic substrates are used and blood sugar is prevented from falling too low; and in the stress response, blood glucose is made available in increased amounts.

Endocrine tissues of the testis and ovary

In the testis, the interstitial (**Leydig**) cells produce the male androgenic steroid testosterone. Their homologues, the interstitial and **theca interna** cells of the ovary, also produce androgens, largely androstenedione. These androgens pass both into the bloodstream and into the germinal tissues.

In the germinal epithelium of the testis, the **Sertoli cells** can convert testosterone both to oestrogen (aromatization) and also to the more powerful androgen dihydrotestosterone. The Sertoli cells also produce a peptide hormone, inhibin, which acts in a negative feedback loop on pituitary secretion of follicle stimulating hormone (FSH).

In the ovary, after puberty, **granulosa cells** of the growing follicles aromatize the androgen produced by the surrounding thecal cells to oestrogen, which is secreted into the circulation in large amounts. The oestrogen exerts many actions, including feedback on the hypothalamo-pituitary endocrine axis, stimulation of growth of the reproductive organs, and conservation of bone mineral. After ovulation, the **corpus luteum** derived from the ovulated follicle produces large amounts of progesterone and some oestrogen. The progesterone also has feedback effects and prepares the reproductive tract for implantation of a fetus. The ovary also produces a number of peptide hormones, including inhibin.

Placenta

The human placenta produces very many hormones. Its syncytiotrophoblast produces very large amounts of progesterone, largely from maternal cholesterol, and smaller amounts of oestrogens (oestradiol and the much less active oestriol) from DHEA-S derived from the fetal and maternal adrenal (see above).

The syncytiotrophoblast also produces variants of many of the peptide hormones produced by the definitive endocrine glands, particularly the pituitary. Chief among these are human chorionic gonadotrophin (hCG), which prevents regression of the corpus luteum in early pregnancy, and human placental lactogen (hPL), a hormone with both lactogenic and growth hormone-like activity which appears to have a major glucose-sparing effect on the metabolism of the mother and corticotrophin-releasing hormone (CRH).

Kidneys

The kidneys are involved in the production of three hormones, but in very different ways.

Erythropoietin is a glycoprotein hormone which stimulates the production of erythrocytes from their precursors and is produced by cells lying between the tubules of the renal cortex (and by the liver in fetuses). Its production is stimulated by low oxygen concentration in the renal circulation. Therefore, the anaemia caused by poor red cell production in renal failure can be treated by administration of erythropoietin. Also, erythropoietin can be abused by athletes seeking greater oxygen-carrying power in their blood.

Specialized juxtaglomerular cells in the walls of the afferent arterioles supplying renal glomeruli secrete the enzyme renin when stimulated by either sympathetic nerves supplying the juxtaglomerular apparatus or low sodium in the distal nephron. The renin acts on angiotensinogen, a plasma protein secreted by the liver, to produce angiotensin I. This is then converted by angiotensin converting enzyme, largely in the lungs, to the active hormone angiotensin II. Angiotensin II stimulates thirst and the release of the adrenal mineralocorticoid aldosterone which causes sodium retention by the kidney and gut. It can also, in high concentration, constrict certain vascular beds—the effect that gave angiotensin its name.

The kidneys also produce the enzyme 1α-hydroxylase, which acts on the inactive 25-hydroxyvitamin D_3 to produce the active hormone 1,25-dihydroxyvitamin D_3, which stimulates the uptake and conservation of calcium and phosphate. The enzyme is activated by parathyroid hormone in response to low plasma calcium concentrations. Renal failure is therefore also associated with loss of body calcium and weak bones.

Gastrointestinal tract

The endocrine cells of the gastrointestinal tract are derived from gut endoderm and occur as isolated cells in the alimentary epithelium, situated mostly in the base of the epithelial glands. They produce a large number of different hormones, including:

- gastrin—a peptide secreted by the gastric antrum, which stimulates secretion of gastric acid and initiates protein digestion;
- histamine—an amine produced by the body of the stomach, which is an essential intermediary in the stimulation of gastric acid secretion;
- secretin—a peptide, produced by the duodenum, which stimulates secretion of alkaline secretions from the liver and pancreas to neutralize gastric acid;
- cholecystokinin—a duodenal peptide, which stimulates secretion of pancreatic enzymes and causes the gall bladder to contract and eject bile into the duo-denum to emulsify the fat for more efficient digestion; and many other peptides for which an endocrinology text should be consulted.

Diabetes mellitus, insulin treatment, and islet transplantation

Diabetes mellitus is a condition in which, because insulin is either not secreted (type 1 diabetes is caused by autoimmune destruction of the islets) or fails to act properly (there is resistance to the action of insulin in type 2, maturity-onset diabetes), glucose cannot enter most tissues, but accumulates in the blood, and is lost in the urine. Many diabetics require treatment by administration of insulin. If this is injected systemically, the peripheral tissues and the liver receive the same concentration of insulin, which is not appropriate for their metabolic demands.

For this reason, attempts are being made to transplant donor islet tissue, primarily by seeding islets into the hepatic portal circulation so that they embed in the liver. In this position, the cells can respond to blood-borne metabolic and hormonal signals and deliver insulin at high concentration direct to the hepatocytes.

Endocrine tumours of the pancreas

The most common endocrine tumours in the pancreas not surprisingly secrete insulin (insulinoma) and cause hypoglycaemic attacks. Other endocrine tumours of the pancreas produce gastrin, which is naturally produced in the fetal islets. Uncontrolled secretion of gastrin causes unwanted secretion of gastric acid and peptic ulceration.

Local hormones

In addition to these hormones produced by particular organs, nearly all tissues produce a range of 'local hormones' such as prostaglandins, growth factors, and other intercellular signals. These are not considered here, but are important in the development of the tissues and organs, the interaction of the organs with the autonomic nervous system, and in the integrated function of the organs.

Questions and answers

Qu. A *Aberrant endocrine tissue derived from endocrine glands normally located in the neck may also be found in the anterior mediastinum. Which glands, and why does this occur?*

Answer The superior and inferior parts of parathyroid glands develop from endodermal tissue of the fourth and third pharyngeal pouches, respectively. Normally, they become attached to the posterior aspect of the thyroid gland in the neck. However, the thymus gland also develops in part from the third pharyngeal pouch and this normally descends into the anterior mediastinum of the thorax. The inferior parathyroid gland may descend with the thymus into the thorax. The developing thyroid gland also migrates caudally from its origin as a diverticulum in the floor of the developing mouth. It normally passes downward from the tongue, and crosses in front of the hyoid bone to become attached to the anterior aspect of the thyroid cartilage and upper rings of the trachea. Occasionally it migrates too far caudally to reach the anterior mediastinum. If it is positioned immediately in the narrow space behind the manubrium in the inlet to the thorax and enlarges there, the trachea can be compressed by the thyroid tissue, causing respiratory obstruction.

Qu. B *From where do suprarenal medullary cells derive preganglionic fibres? What is the principal neurotransmitter at the synapse between preganglionic fibres and the adrenal medullary cells?*

Answer Cells of the adrenal medulla receive preganglionic sympathetic fibres from the thoracic splanchnic nerves. The cell bodies of these neurons lie in the lateral horns of the thoracic spinal cord. Their transmitter is acetylcholine.

Anatomical bases of some neural reflexes of the thorax, abdomen, and pelvis

Anatomical bases of some neural reflexes of the thorax, abdomen, and pelvis

Reflexes help maintain homeostasis and support our interactions with our internal and external environments. This chapter summarizes the neural pathways involved in selected reflexes of the thorax and abdomen. Thoracic and abdominal viscera are controlled by many reflexes which involve the interaction of the autonomic and somatic nervous systems and the endocrine system. The text is purposely brief and emphasizes the basic components of a reflex. These are:

- a **sensory** (afferent) **neuron** which transmits information from the internal or external environment to the central nervous system;
- an interconnection in the central nervous system which usually involves one or more neurons (**interneurons**) other than the sensory and motor neuron;
- a **motor** (efferent) **neuron** (somatic or autonomic) which passes from the central nervous system to the effector tissue to evoke the reflex response.

The least anatomically complex reflexes involve a direct connection between the sensory and motor neuron (monosynaptic reflex; the 'stretch' reflex, e.g. knee jerk). However, most reflexes also involve one or more central interneurons interposed between the afferent and the efferent neuron (polysynaptic reflex). Such interneurons lie at, below, and above the level at which the afferent neuron enters the central nervous system. The original sensory stimulus therefore spreads locally and/or to higher centres, thereby enabling the original information to be processed and integrated before the motor response is stimulated. Motor responses are usually stereotyped and may be common to a number of different sensory stimuli. Descending influences from higher centres can have a marked effect on the sensitivity of reflexes. The presence and quality of the reflex responses therefore provides an indication of both the integrity of the reflex arc and also the degree of excitation of that arc from other centres.

Neurons involved in the reflexes

Sensory neurons convey sensation from both somatic tissues and viscera. Their peripheral axons pass toward the central nervous system; their (unipolar) cell bodies are located in the dorsal root ganglia; and their central processes pass through the dorsal roots into the spinal cord. In the cord they make synapses within their own and adjacent segments of the cord, and also project to higher levels.

Somatic motor neurons have their cell bodies in the ventral horn of the spinal cord. Their long axons emerge through the ventral nerve roots and pass with spinal nerves and their branches to terminate in their target muscles.

Preganglionic sympathetic neurons have cell bodies in a lateral horn of grey matter. This extends as a column of cells between T1 and L2 segments of the cord (thoraco-lumbar outflow). Their axons pass via the ventral nerve roots to ganglia of the paravertebral sympathetic chains, or to the midline autonomic ganglionic plexuses of the thorax, abdomen, and pelvis (e.g. cardiac, pulmonary, coeliac, pelvic) where they synapse on ganglionic neurons. These **ganglion cells** project relatively long unmyelinated (postganglionic) axons, which either join spinal nerves for distribution to the body wall, or pass through the midline autonomic plexuses for distribution to their target organs.

Preganglionic parasympathetic neurons which supply the thorax and abdomen are found in the dorsal motor nucleus of the vagus (X) nerve and in S2, 3 and 4 segments of the cord (craniosacral outflow). (Those supplying the head are associated with cranial nerves III, VII, IX, X; see Vol. 3). Their long axons project to groups of ganglionic neurons which lie within the walls of thoracic and abdominal viscera. From these ganglion cells short axons pass to their targets on the smooth muscle and glandular cells of the viscera.

The diagrams in this chapter have been standardized as far as possible. Somatic components are coloured yellow, sympathetic components beige and parasympathetic components orange. The pelvic plexus is represented as a square box.

Cough reflex (8.1)

- **Sensory**. Irritant stimuli to the trachea or bronchi reach the central nervous system through visceral afferent fibres which run with efferent fibres of the **sympathetic** system and the **vagus** nerve which supply the lung.

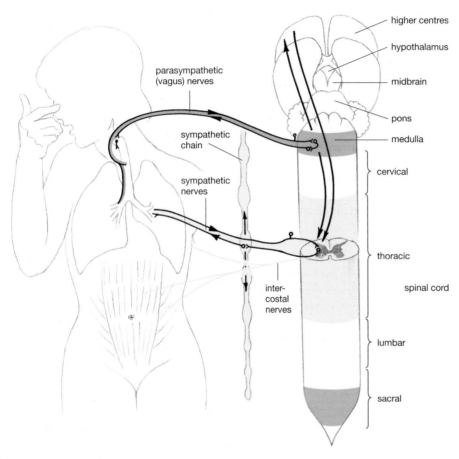

8.1 Cough reflex.

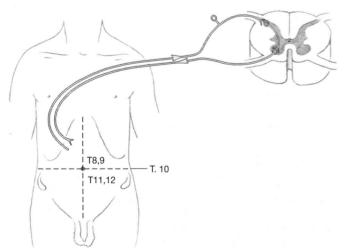

8.2 Abdominal reflex.

- The impulse spreads through interneurons in the **brainstem** and cord, and the sensitivity of the reflex is markedly affected by higher centres.
- **Motor**. Thoracic motor neurons send fibres in the **intercostal nerves** to cause contraction of intercostal and abdominal muscles. At the same time motor fibres in the **vagus** nerve close the airway at the glottis. The intrathoracic pressure is first increased, then when the glottis suddenly opens, the change in airway pressure can dislodge mucus or other material which has entered the upper respiratory tract and expel it into the mouth.

Hiccoughing reflex

- **Sensory**. Irritation of the diaphragm reaches the central nervous system through afferent fibres of the **phrenic nerve (C4)**.
- The information spreads through interneurons in the cord.
- **Motor**. Neurons of the **phrenic nerve (C4)**, cause a brief contraction of the diaphragm and 'catching' of the breath. The contractions are often repeated until the reflex circuit is broken by 'holding the breath' (deep respiration against a closed glottis) or other respiratory manoeuvres.

Abdominal skin reflexes (8.2)

- **Sensory**. Stimuli (e.g. from stroking the skin) to the anterior abdominal wall in each of its quadrants are carried by afferent fibres of **intercostal nerves (T6–T12, L1)** to the central nervous system.
- The stimulus spreads through interneurons in the thoracic cord. The reflex is much influenced by fibres descending from higher centres.
- **Motor**. Motor neurons of the lower **intercostal nerves** activate a localized contraction of muscles in the same abdominal segment as the original stimulus.

Cremasteric reflex (8.3)

- **Sensory**. Stimuli (as above) to the skin of the medial aspect of the thigh are carried by afferent fibres of the **ilio-inguinal nerve (L1)** to the central nervous system.
- The stimulus spreads through interneurons at L1 level of the spinal cord.
- **Motor**. Motor neurons of L1 spinal segment give fibres to the **genitofemoral nerve (L1, 2)**. These cause contraction of the cremaster muscle which elevates the testis on the same side.

Vomiting reflex (8.4)

- **Sensory**. Many different stimuli, such as irritation of the stomach (**vagus**), severe pain, fear, obnoxious smells, or disturbance of the semicircular canals (**vestibular nerve**), can trigger vomiting. The information is transmitted by afferent nerves to appropriate centres within the central nervous system.

- The stimulus spreads, in particular to a 'vomiting centre' located in the medulla of the brainstem.
- **Motor.** Efferent **sympathetic nerves** are activated. This, with increased circulating adrenaline, causes nausea, increased salivation, slower and deeper breathing, perspiration, and constriction of cutaneous blood vessels. At this stage, higher centres can inhibit vomiting for a variable length of time. However, if the afferent stimuli increase, motor fibres of **lower intercostal** and **phrenic nerves** stimulate contractions, respectively, of anterior abdominal wall muscles and of the diaphragm. This raises intra-abdominal pressure and causes stomach contents to be ejected through the oesophagus and mouth.

Vomiting is accompanied by other important **motor** effects. The laryngeal opening closes (**vagus nerve**) to prevent vomit from entering the trachea and lungs. A ring of muscle around the soft palate at the junction of the naso- and oropharynx contracts to protect the nasal airway, but this is often less effective than in the swallowing reflex (see Vol. 3).

Micturition reflex (8.5)

While urine is collecting in the bladder, emptying is prevented by inhibition of the detrusor muscle and by closure of the bladder sphincters (sympathetic nerves and pudendal nerve S2, 3, 4).

- **Sensory.** Stimuli arise from stretch receptors in the bladder wall as it fills with urine. Afferent fibres passing with both **sympathetic** and **parasympathetic** nerves conduct the information to the central nervous system.
- A 'micturition centre' is present in the lower lumbar and sacral parts of the spinal cord. A centre in the brainstem pons and an area of 'motor' cortex are also involved. When the volume of urine becomes sufficiently large, reflex bladder emptying can be initiated.
- **Motor.** Parasympathetic fibres to detrusor cause contractions of the bladder wall. At the same time, the voluntary sphincter is relaxed.

'Fullness of the bladder' is perceived well before maximal filling. Mild reflex contractions of the bladder wall occur, but the micturition reflex can be inhibited by higher neural centres until a socially convenient place is found to void the urine. Eventually, when the bladder is very full, the reflex can overcome any inhibition by higher neural centres.

Micturition is started voluntarily, by relaxing the external urinary sphincter (**pudendal nerve S2, 3, 4**) and, if necessary, raising intra-abdominal pressure. This is accompanied largely by excitation of efferent **parasympathetic** fibres (**pelvic splanchnic nerves S2, 3, 4**) supplying the detrusor.

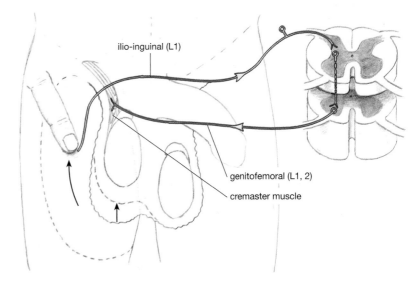

ilio-inguinal (L1)

genitofemoral (L1, 2)

cremaster muscle

8.3 Cremasteric reflex.

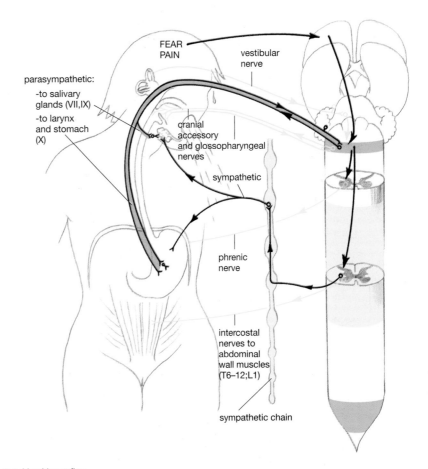

FEAR
PAIN

vestibular nerve

parasympathetic:
-to salivary glands (VII,IX)
-to larynx and stomach (X)

cranial accessory and glossopharyngeal nerves

sympathetic

phrenic nerve

intercostal nerves to abdominal wall muscles (T6–12;L1)

sympathetic chain

8.4 Vomiting reflex.

In babies and infants the reflex involving the bladder and spinal cord is apparently all that is required for voiding of urine. However, as age increases, the integrity of a pontine centre becomes essential for normal functioning of the micturition reflex.

Transection of the spinal cord above sacral level abolishes the influence of the pontine centres (and other higher centres). After such an injury the bladder is at first unable to empty (spinal shock). Later, partial emptying of the bladder occurs via the spinal reflex whenever the bladder becomes sufficiently full (automatic bladder). Therefore, after such an injury and during the initial period when the bladder must be artificially drained, it should not be permitted to remain empty, because it would then become permanently contracted and unable to refill to a normal extent when the spinal reflex becomes reactivated

The recovered reflex is usually triggered by lesser volumes of urine than is normal and, because the bladder never empties fully, the residual pool of urine is prone to become infected.. Provided that the spinal autonomic innervation to the bladder remains intact, patients can learn to trigger the spinal reflex, at a socially convenient time and place, by stimulating an appropriate segmental area of skin (suprapubic).

Defaecation reflex (8.6)

Faeces accumulate in the descending and sigmoid colon, and are intermittently moved into the rectum. The internal anal sphincter is tonically active (**sympathetic**).

- **Sensory**. When faeces stretch receptors in the wall of the rectum sensory fibres, passing largely with **parasympathetic** nerves of the pelvic plexus, conduct the information to the central nervous system. This stimulates the desire to defaecate and sympathetic fibres relax the internal anal sphincter.
- An integrative centre for defaecation is located in the sacral spinal cord. When 'fullness of the bowel' is perceived and mild reflex contractions of the rectum occur, defaecation can be inhibited by higher neural centres activating the voluntary external anal sphincter until such time as a socially convenient place can be found.
- **Motor**. **Parasympathetic** neurons of the sacral spinal cord cause expulsive contractions of the rectal and colonic smooth muscle. Defaecation is started voluntarily by relaxing the external anal sphincter (**pudendal nerve S2, 3, 4**) and, if necessary, increasing the intra-abdominal pressure. This is normally accompanied by inhibition of efferent **sympathetic** fibres to the internal anal sphincter. The pelvic floor is lowered and the entire lower bowel evacuated in a concerted action.

Transection of the spinal cord above sacral levels abolishes the voluntary activity that assists the defaecation reflex. However, as with the bladder, if the autonomic innervation of the lower bowel remains intact, reflex defaecation will occur as the rectum fills, and the patient can be trained to initiate the reflex by stimulating an appropriate area of skin.

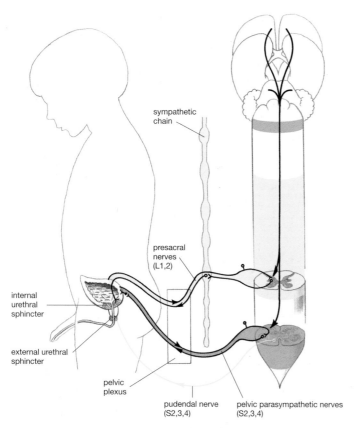

8.5 Micturition reflex.

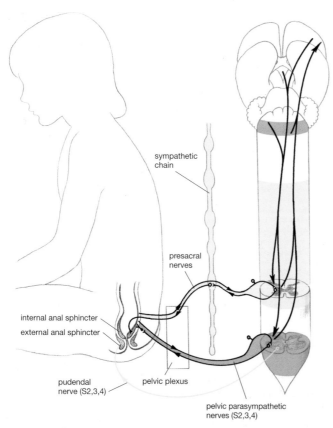

8.6 Defaecation reflex.

Genital reflexes (8.7)

- **Sensory.** In men and women stimuli from higher centres (e.g. visual, olfactory, auditory, and psychological), from cutaneous erogenous zones (e.g. nipple, back of neck), and from the genitalia reach the central nervous system via afferent fibres of **somatic**, **sympathetic**, and **parasympathetic** nerves.
- Centres are present in the sacral cord (for erection, ejaculation), and in the lower lumbar cord (for emission).
- **Motor.** Increased activity of efferent **pelvic parasympathetic** fibres inhibit constriction of helicine arteries of the penis or clitoris, thus enhancing blood flow to the erectile tissue and causing *erection*. They also stimulate secretion from genital mucous glands, to provide lubrication.

In **men**, increasing afferent stimuli lead to excitation of efferent **sympathetic** fibres which causes contraction of the smooth muscle of the vas deferens, prostate gland, and seminal vesicles, which results in *emission* of sperm and seminal fluid into the bulb of the urethra. At the same time the internal urinary sphincter contracts, preventing back-flow of semen into the bladder.

Efferent fibres of the **pudendal nerve** (somatic) then stimulate contraction of the (striated) muscles of the penis (bulbospongiosus and ischiocavernosus) causing *ejaculation* of the urethral contents.

Damage to the sympathetic pelvic splanchnic nerves results in impotence. If the lesion is above the sacral level of the cord, erection can sometimes occur either spontaneously or after mechanical stimulation, although coordination with the processes of emission is unlikely.

Orgasm in men starts during emission and ends in ejaculation.

In women sexual stimulation causes the clitoris to become erect. Rhythmic contractions of levator ani, many fibres of which pass around the vagina, occur during the phases equivalent to emission and ejaculation in males.

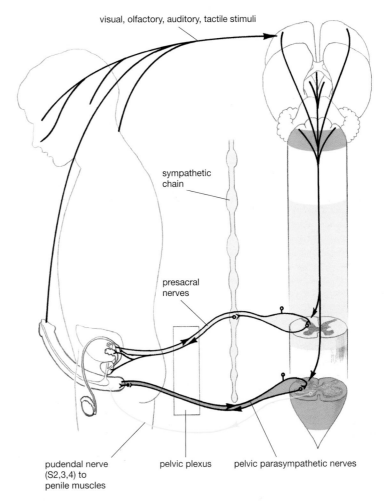

8.7 Genital reflexes.

Index